D1631599

THE CANALS OF SCOTLAND

THE CANALS OF THE BRITISH ISLES

EDITED BY CHARLES HADFIELD

1. *British Canals. An illustrated history.* By Charles Hadfield
2. *The Canals of South West England.* By Charles Hadfield
3. *The Canals of South Wales and the Border.* By Charles Hadfield
4. *The Canals of the North of Ireland.* By W. A. McCutcheon
5. *The Canals of the East Midlands (including part of London).* By Charles Hadfield
6. *The Canals of the West Midlands.* By Charles Hadfield
7. *The Canals of the South of Ireland.* By V. T. H. and D. R. Delany
8. *The Canals of Scotland.* By Jean Lindsay
9. *Waterways to Stratford.* By Charles Hadfield and John Norris

in preparation

10. *The Canals of North West England.* By Charles Hadfield and Gordon Biddle
11. *The Canals of South and South East England.* By Charles Hadfield
12. *The Canals of North East England.* By Charles Hadfield and Gordon Biddle
13. *The Canals of Eastern England.* By J. H. Boyes

THE CANALS OF SCOTLAND

by

Jean Lindsay

WITH PLATES AND MAPS

DAVID & CHARLES: NEWTON ABBOT

7153 4240 1

Printed in Great Britain by
Latimer Trend & Company Limited Plymouth
for David & Charles (Publishers) Limited
South Devon House Newton Abbot Devon

For
CORA and JOHN

CONTENTS

ILLUSTRATIONS

PLATES

MAPS

PREFACE

THE feature which most clearly distinguishes the history of inland navigation in Scotland from the history of inland navigation in England and Ireland is the fact that the major Scottish rivers are so shallow, so swift-flowing and so broken by falls that navigation has never been possible on them for more than a few miles from the sea. Whereas in England and Ireland, therefore, the construction of canals in the eighteenth and nineteenth centuries provided final links in systems of inland communication which were already extensive and constantly-used, the inauguration of the canal era in Scotland was the beginning of inland navigation in that country. The great waterway of pre-industrial Scotland was the sea; and since such navigable estuaries as those of the Clyde, the Forth and the Tay penetrated far into the country and gave sea-going vessels access to most of the major centres of population, the sea went far towards satisfying the country's transport requirements. Even in the time of the great waterway engineers, the improvement of navigation in these estuaries and in the lower reaches of their rivers made a more important contribution to Scottish economic development than did the construction of new waterways within the country; and both these developments were conceived in the first place as means of expediting coastal navigation and extending it inland. The history of inland navigation in Scotland does not begin, then, until the later eighteenth century; and one can satisfactorily define it as a subject distinct from the history of coastal navigation only by deeming it co-extensive with the history of the Scottish canals.

As the canals of England and Ireland united existing river-navigations with one another, so the more important of the Scottish canals were created to provide new links in the established waterway system represented by the sea, the estuaries and the larger fresh-water lochs. At the same time, the desire for easier communication between major centres of mineral and agricultural production and major centres of population, which stimulated many of the most important of the English canals, played its part in developing the secondary branches of the Scottish canal system;

and the need for viable routes from mines and quarries to harbours which could be reached by coastal steamers was also a contributing factor. The canals which were built to satisfy these varied requirements were not wholly independent of one another—they provided through routes, for example, between Paisley and Edinburgh and between Leith and Inverness—but they did not make up a canal-network like that which existed in the Midlands. None of Scotland's seven major canals—not even the Union—was originally projected as an inland waterway to link with another inland waterway; and a surprising number of the minor canals were land-locked, communicating neither with other waterways nor with the sea. While general tendencies in Scotland's economic development exerted similar influences, therefore, on all the canals, it is possible and necessary in examining the Scottish canals, as it is not in examining the canals of England or even Ireland, to consider each of them as having an independent history. The first seven chapters of this book are thus concerned with the seven major canals; and such subjects as the Forth & Clyde water-supply and the Glasgow–Inverness passenger-trade are dealt with where they first become relevant. In the final chapter I have given a brief account of the smaller canals and of the more interesting of those proposals for the improvement of inland navigation which were never put into effect.

JEAN LINDSAY

CHAPTER I

The Forth & Clyde Canal

ON 24 September 1726 Alexander Gordon, who had been commissioned by the government to survey the route for a canal to transport goods between the Firths of Forth and Clyde, wrote from Kilsyth to the Barons of the Exchequer in Edinburgh: 'After much bad weather and incredible fatigue I am at last half-way between the two seas of Britain. As for the project in general the facility of effecting it greatly surpasses the idea which I had of it.' Gordon, who is chiefly remembered now as the author of *Itinerarium Septentrionale*, a survey of Roman remains in Scotland and the North of England, regarded Dullatur Bog and the River Kelvin as the only serious obstacles to the construction of a canal linking the two firths. According to Knox's *View of the British Empire*, the notion of such a canal had previously been mooted by Charles II, who proposed to open a passage for warships at a cost of £500,000; and in Gordon's time it received the support of Daniel Defoe, who believed it would be a simple matter 'to make a navigation from the Forth to the Clyde, joining the two seas'. In 1741 William Wishart referred to Gordon's work and expressed the wish for 'a new and exact survey'; and a letter written in the same year by William Adam, who had examined the ground in Gordon's day, reveals that the magistrates of Edinburgh and Glasgow were thinking of 'setting the old project on foot' at an expense which Adam estimated at £200,000.[1]

In 1760 the Earl of Chatham suggested that the scheme might be carried out at the public expense; but he resigned soon afterwards, and the idea was dropped. Two years later, Lord Napier of Merchiston employed Robert Mackell and James Murray to make a survey from the River Carron at Abbotshaugh to the Clyde at Yoker Burn, about 5 miles below Glasgow. As a result of this report and of Lord Kames' enthusiasm for the scheme, the Board of Trustees for the Encouragement of Fisheries, Manufactures and Improvements in Scotland, set up and financed by the state in

1727, requested the Yorkshire engineer John Smeaton to make another survey in 1763. Smeaton, who had completed the Eddystone Lighthouse in 1759, made his report in 1764, suggesting two possible routes. One of these was from the River Carron by the valley of the Bonny through Dullatur Bog into the valley of the Kelvin and then into the Clyde at Yoker Burn; the other followed the Forth for some miles above Stirling, crossed the Bog of Bollat into the River Endrick, and then proceeded to Loch Lomond and by the River Leven into the Clyde at Dumbarton. The Loch Lomond scheme was the more expensive and less direct of the two, and was never seriously pursued. Smeaton estimated the cost of a canal 5 ft deep along the 27 miles of the Carron route at £78,970; and in an appendix he ranked this proposed canal with 'the noblest work of the kind that ever has been executed, the Canal Royal of Languedoc'.[2]

Smeaton's report was debated for more than two years. Many hoped that public funds would be used to put it into effect; others hoped that the scheme would collapse. The tobacco-merchants of the west, who wanted such a canal as would facilitate the re-exporting of American tobacco from Glasgow to the continent, were indignant at the by-passing of their city in Smeaton's plan; and in December 1766 they commissioned Robert Mackell to 'examine whether there might not be found some track, which, though it was higher, might be shorter, and which should fall into the Clyde nearer Glasgow'. Mackell drew up proposals for a canal 7 miles shorter than Smeaton's, which would pass through the North Lanarkshire coalfield and have a 240-ft summit near the Bishop Loch; but this scheme was rejected as too expensive, and Mackell and Watt were asked instead to investigate the possibility of adapting Smeaton's route so that the canal could finish near the Broomielaw. Initially, it was suggested that this canal should be 18 ft wide at surface and 3 ft deep; but when the promoters came to draw up their bill the surface width was given as 24 ft and the depth as 4 ft. This bill, which assumed a total cost for the canal of about £30,000, received the support of the Carron Iron Company because the eastern entry was fixed at Carronshore; it was presented by Lord Frederick Campbell on 16 March 1767, and given its second reading four days later. The merchants of Bo'ness, wishing to preserve their town's importance as a port on the Forth, petitioned for a canal from Carronshore to Bo'ness of the same width and depth as that from Glasgow to Carronshore; and this proposal was incorporated in the bill.[3]

These projects aroused much excitement in the press, the sup-

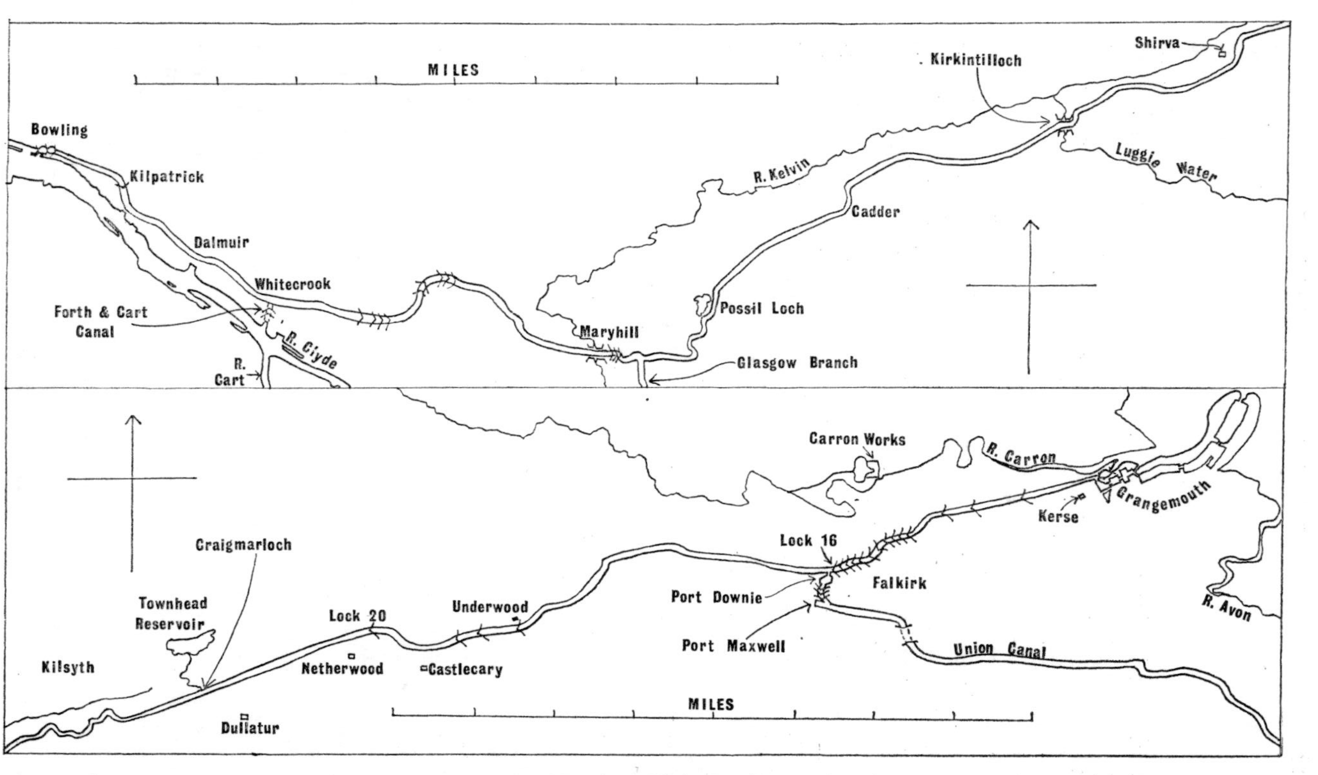

1. Forth & Clyde Canal

porters of the Glasgow scheme arguing that a larger canal would not cheapen transport costs between the Forth and the Clyde, and maintaining that nine-tenths of the goods to be carried through the canal would be goods coming from or going to Glasgow. Mackell and Watt, having surveyed two possible routes for the small canal, produced an estimate of £50,000 for the more northerly and less expensive of them. The traders of the Forth were alarmed by the Glasgow bill's success, and on 3 April the merchants and landowners of Midlothian met in Edinburgh and resolved to present a petition against it. The petitioners, who included the Lord Provost, Magistrates and Town Council of Edinburgh, asked Parliament to postpone action on the 'partial and local' scheme under consideration so that plans could be drawn up for a 'proper canal' which would serve 'national and universal' interests.[4]

Correspondence in the press became heated, and the latent hostility between Glasgow and Edinburgh was brought into the open. One writer referred ironically to the 'little despicable scheme of the Glasgow merchants' for 'a ditch, a gutter, a mere puddle' which would serve the purposes of trade but not those of 'magnificence and national honour', and continued: 'What is commerce to the city of Edinburgh? Edinburgh, Sir, is the metropolis of this ancient kingdom, the seat of Law, the rendezvous of Politeness, the abode of Taste, and the winter quarters of all our nobility who cannot afford to live in London; and for these and other reasons equally cogent Edinburgh ought to have the lead upon all occasions. The fools of the west must wait for the Wise Men of the East.' The petition against the small canal was supported, however, by the counties of Aberdeen, Banff, Elgin, Fife, Haddington, Linlithgow, Perth, Selkirk and Stirling; and on 1 May 1767 the subscription for a 'great canal' was opened in London. In a short time over £100,000 had been subscribed; and on 13 May the supporters of the small canal agreed to drop their bill in return for the payment of £1,200 compensation to the Glasgow–Carron subscribers and £300 to the Bo'ness subscribers and the inclusion in the new bill of provision for collateral cuts to Glasgow and Bo'ness. The triumphant subscribers for the 'great canal' held their first meeting in London on 27 May, with the Duke of Queensberry in the chair. The committee, which included the former commissary-general and contractor to the army Sir Lawrence Dundas of Kerse, ordered a call of 5 per cent on subscriptions and instructed Smeaton to plan a canal for ships of 60 tons burden.[5]

Smeaton's second report came out in October, and referred only to the Carron route. In his introduction, Smeaton explained that his original scheme had not been intended for sea-going vessels, but that those who were 'versed in trade' believed that a canal of 7 to 10 ft in depth would be a greater benefit to the public. This canal was to have a branch to Glasgow and run from a western entry at Dalmuir to an eastern entry near the mouth of the Carron on the north-east side of a farm called 'the Heuck'. At the eastern end there was a lack of shelter for sea-going vessels delivering their cargoes, and to alleviate this difficulty Smeaton proposed an additional entry to the canal 'from about the Grangeburnfoot'. The problem at the western end was the state of the Clyde, which was too shallow above Dalmuir for vessels of more than 4 ft draught. The estimate for a canal 7 ft deep was £147,337; but Smeaton did not expect his report to be the last word, for he added, 'There must be a degree of latitude in an affair of such great consequence for second thoughts and improvements.'[6]

Smeaton's recommendations were approved by the Convention of Royal Burghs, and in December Parliament was petitioned for leave to bring in a bill. Thomas Dundas, the son of Sir Lawrence, told Parliament that the cut from the mouth of the Carron to 'Dalmuir Burnfoot' and the collateral cut from 'Three Part Miln' to Glasgow would not only 'open an easy communication between the Firths of Forth and Clyde' but also be 'of great advantage to the kingdom in general by reducing the price of land carriage'. Leave to bring in the bill was granted; and it had its second reading in January 1768. At a committee meeting in Edinburgh the following month George Chalmers, who had been sent to London to help the bill through, stated that Charles Gascoigne, the manager of the Carron Company, had agreed to the bill only when three of the subscribers had undertaken that the canal would be linked both with the Carron works at 'Stenhouse Miln Dam' and with a convenient point further down the river. Chalmers also reported opposition from the port of Bo'ness; and when the bill had its third reading it contained an amendment 'for making a navigable cut or canal of communication from the port and harbour of Bo'ness to join the canal at or near the place where it falls into the Firth of Forth'. There was pressure, too, for a depth of 10 ft, particularly from those who regarded the canal as a 'general navigation' likely to benefit the British Isles as a whole. After the Act had been passed, the merchants of Edinburgh petitioned the Convention of Royal Burghs to apply for money from the forfeited estates so that the canal could be made deep enough to take sloops drawing 8 or 10 ft.[7]

The preamble to the Act, which received the royal assent on 8

March 1768, claimed that the canal would help 'the improvement of the adjacent lands, the relief of the poor, and the preservation of the public roads', as well as facilitating trade between the two firths. The proprietors included the Dukes of Bedford, Buccleugh, Gordon and Queensberry, the Earls of Morton, Abercorn, Rosebery, Panmure, Fife and Catherlough, and the Lord Provosts of Glasgow and Edinburgh. They were empowered to make a canal 7 ft deep from near the mouth of the Carron by way of Bainsford, Bonny Mill, Dullatur Bog, Inchbelly Bridge, Cadder Bridge and St Germain's Loch to the Clyde near 'Dalmuir Burnfoot', and a collateral cut from the main canal near Blairdardie through Partick and across the Kelvin to Glasgow. The company was authorized to raise £150,000 in shares of £100, and if that was insufficient to raise a further £50,000 by a new subscription. Power was also given for the construction of a cut to Bo'ness; but this was to be a separate undertaking financed by the proprietors of the Bo'ness Company. The proprietors of the Forth & Clyde were authorized to draw water from the Rivers Carron, Endrick and Kelvin, and from all rivers and lochs within 10 miles of the canal, provided they did not diminish the water-supply to the mills on the Carron and the Kelvin. Tolls were not to exceed 2d per ton per mile, and ironstone and limestone were to be carried at half and quarter of that rate respectively. Dividends were not to exceed 10 per cent.[8]

The first General Meeting was held at St Alban's Tavern in London on 14 March 1768, with Queensberry in the chair. Smeaton was appointed 'Head Engineer' at £500 per annum, and Robert Mackell 'Sub-Engineer' at £315 per annum. Smeaton proposed that the work should be divided into three sections, with a surveyor and foreman responsible for each; but this was rejected on grounds of expense. Authority was given for a payment of £1,500 to the subscribers for the small canal; and the agreement with Gascoigne for a forked cut from the canal to two points on the Carron was confirmed. The early committee meetings were held at the Exchange Coffee House in Edinburgh, and at one of these it was decided that digging should begin 'at or near the west side of the Grange Burn'. Mackell was ordered to buy tools and engage workmen: the wages of 'common labourers' were not to exceed 10d per day 'except in extraordinary cases', and the overseer was to be paid 'the lowest weekly wages' possible. Smeaton recommended the purchase of beech from Sussex, which he said was the 'best and cheapest', and the importation from Italy of 50 to 100 tons of pozzuolana cement; and he also requested permission to send for a 'mason and digger' who had been employed on the

Calder Navigation. Digging was begun on 10 June, Sir Lawrence Dundas removing the first spadeful and distributing 5 guineas among the workmen.[9]

There was last-minute opposition, however, from the Carron Company and others. James Brindley, Thomas Yeoman and John Golborne were commissioned to make a survey of the area between the Forth and the Clyde, and their reports were published in October. Brindley proposed a canal 4 ft deep with lock dues of 1½d per ton, and recommended that the River Carron should be diverted through the neck of land south of the Heuck Farm; Golborne suggested that the canal should be re-routed to finish nearer Glasgow. These comments provoked indignant reactions from the newly-established company. William Pulteney, learning of them before their publication, declared that they were the result either of a desire to please the Carron Company or of a confusion between the function of a sea-to-sea canal and that of 'an English inland navigation', and alleged that Brindley had admitted in conversation 'that he was no judge of the matter, not knowing even the very goods that were to pass'. Smeaton, regarding the publication of the reports as an unwarranted attempt to stop the progress of a legally-sanctioned design, pointed out in a detailed reply that the cutting of the proposed channel would involve dredging 'Reay's ford', which would be a very expensive operation, and that an entry nearer to Glasgow would be impracticable because of the shallowness of the Clyde.[10]

Those who supported Brindley's scheme for the cutting of the neck and the consequent improvement of the Carron were doubtless aware of the advantages this would bring to the Carron Shipping Company, which had been established in 1765 by Gascoigne and Francis Garbett and was now engaged on trade between Carronshore and the ports of the Baltic. Smeaton's plan, on the other hand, offered the prospect of private advantage to Sir Lawrence Dundas, who owned the estate of Kerse across which the Grangeburnfoot entry would run; and one of the proprietors judged it necessary, in a public address, to deny charges that the committee had been 'biassed in their judgment to favour the interest of a certain great land-proprietor who supported the undertaking from the beginning'. Dundas' enemies, indeed, accused him not merely of wanting to increase the value of his own land but also of being resolved, out of 'implacable enmity and unmerited private resentment', to 'block up the navigation of the river and ruin an iron manufactory' thereon.[11]

A further attempt to modify the plans for the eastern entry was made in November by the Carron Shipping Company, who pointed out in a representation to the canal company that they had improved the river at their own expense and required easy access to the Forth if they were to benefit from this, but that Smeaton's plan would create a situation in which a few ships with goods for the canal could obstruct the navigation of the river. Smeaton, who had been directed by the canal company to concentrate on 'carrying the main design into execution', replied that any further work on the proposed side-cuts or the possible re-routing of the Carron would divert him from his principal task; and no further action was taken on these matters for the time being.[12]

Work on the 'main design' progressed steadily in 1769, contracts being signed for locks, and lighters being built to carry stone from Kinnaird Quarry to 'Sea-Lock'. It was resolved that proprietors in arrears should be charged interest, and that advertisements for American timber should be inserted in the press. In September it was reported that the number of workers was down to 600 'on account of the harvest'—which suggests that the majority had been agricultural labourers. In November, when the first lock was virtually complete, work was held up in order that 'the curious' might inspect some Roman remains which had been discovered on the site of the Antonine Wall near Castlecary. By the beginning of 1770, however, the canal had been cut for a distance of 9 miles from the eastern entry.[13]

In June of that year Mackell proposed a major change in the route westward from Kirkintilloch, suggesting that the canal should be taken further south so as to cross the Kelvin just 2 miles from Glasgow: he claimed that this would save two years of construction time and £35,333 by reducing the length of the Glasgow branch, dispensing with the need for aqueducts over the Endrick, and cutting down on the cost of land. Smeaton was sceptical about these claims; but he approved the new line, and the necessary Act was passed in March 1771. In May 1770 1,048 men were at work on the canal; and by November all the locks east of the summit had been completed. The following summer, four Roman altars were found near Auchindavy: they were presented to Glasgow University, and can still be seen in the Hunterian Museum there. Smeaton had been dissatisfied from the first with the organization of the work, and in June 1771 he made a lengthy speech to the committee complaining of the lack of 'proper officers' to deal with landowners and buy materials. Recognizing that the committee's unwillingness to appoint such officers stemmed from motives of economy, he offered to resign in order

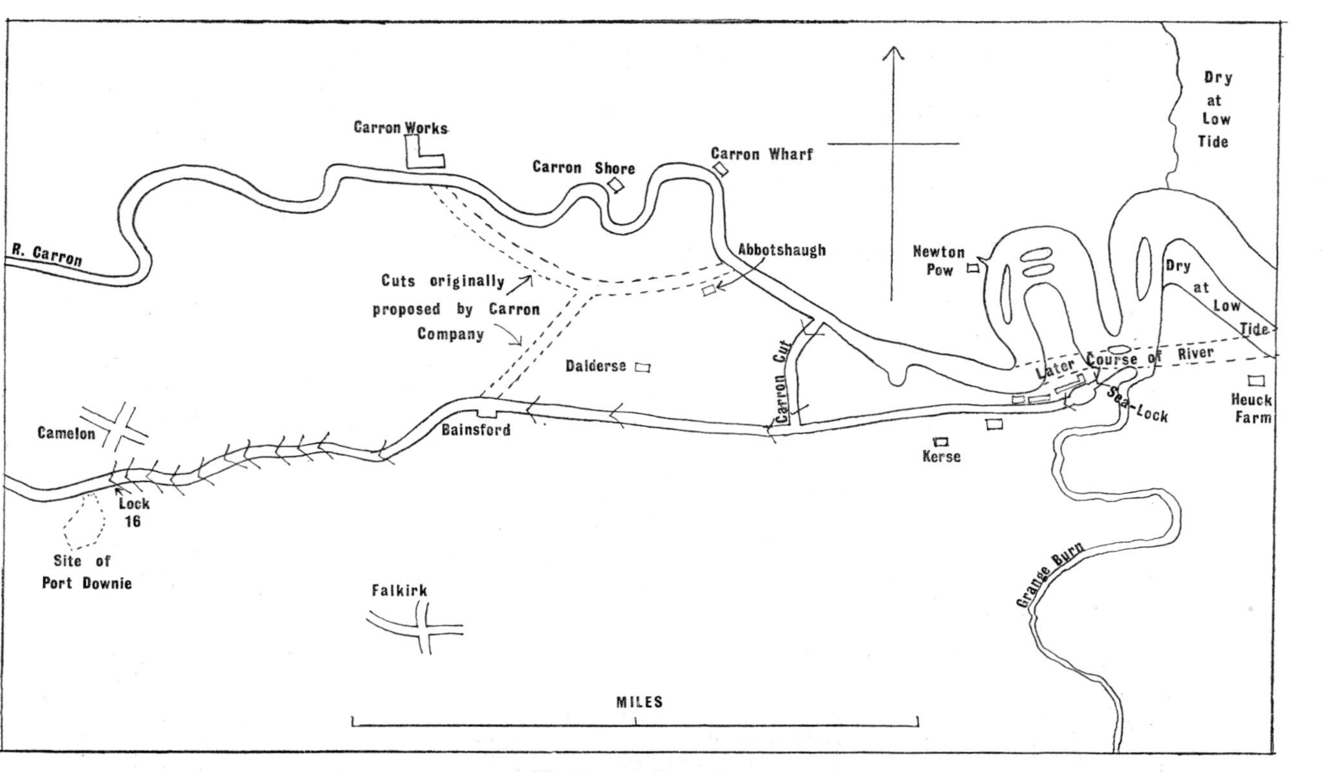

2. The Eastern Entry in 1773

to save them the expense of his salary; but they refused to part with him, and took no action on his proposals for reorganization.[14]

In September Smeaton and Mackell reported on the proposed cut from the canal to the Carron Works, three possible routes for which had been examined by Gascoigne and the canal company's surveyor John Laurie. The cut suggested by Gascoigne was to start from the fourth or Dalderse lock, and was expected to cost £5,725. The shortest and cheapest route, however, was from the fifth or Bainsford lock; and this was the only one for which no further Act of Parliament was required. Smeaton and Mackell recommended that such an Act, if needed, should be obtained at the Carron Company's expense, and pointed out that it would not in any case be 'advantageous' for the canal company to start any branch until separate funds for it were supplied, or to let any subordinate undertaking interfere with overall progress. Reporting at the same time on the main canal, Smeaton warned the company that the cost was exceeding his expectations because of the 'massive and substantial' way in which the work was being carried out.[15]

Gascoigne wrote to Smeaton in November proposing a cut from the Dalderse lock; and when it was suggested that the Carron Company should make this cut at their own expense, he replied that they would be prepared to do this, provided they could thereafter charge the cost to the canal company and take repayment in the form of tolls. The canal company took legal advice on the matter; and at their general meeting in December it was reported that the Carron Company's claims had no legal basis, since the Act did not empower the proprietors to employ their funds on anything other than the main canal and the branch to Glasgow, and that the Carron Company had no right to make a communication linking the canal with the sea even at their own expense, since this would reduce the canal company's income in tolls. In this way, the proprietors successfully avoided carrying out the undertaking which had been given to Gascoigne before the passing of the Act.[16]

By the end of 1771, all the cutting east of the summit had been completed, and work had begun on the dam for the Townhead Reservoir near Kilsyth, which was to cover 70 acres and provide 3,000 lockfuls of water. Early in 1773, quarrels developed in the company, Mackell being accused by some proprietors of 'improper and irregular' behaviour but defended by Smeaton because of his 'great merit in conducting the works'. One of the many arguments concerned the cutting done by the contractors Clegg and Taylor between Dullatur Bog and Kirkintilloch. Mackell claimed that the re-measuring of this had been negligently done; and after a series of angry letters between Mackell and Laurie it was established by means of a further re-measurement supervised by Smeaton that Laurie had in fact overestimated the amount excavated by more than 69,000 cu yd. The committee dismissed Laurie and thanked Mackell for his efforts; but some of its members, having been accused by Mackell of drunken incompetence in the supervision of Laurie's work, must have felt dissatisfied with this conclusion.[17]

In May 1773 Queensberry signed an agreement with the Carron Company for the cutting of one of the loops of the River Carron; but for financial reasons nothing came of this for ten years. The proprietors' credit with the Royal Bank of Scotland had now been reduced to about £500; and the depression resulting from the failure of the Ayr Bank made it difficult to obtain the additional funds necessary for the continuation of the work. Smeaton resigned his post in August, and received the unanimous thanks of the company for the 'very masterly manner' in which he had planned the work; he had indeed shown not only engineering skill in dealing with such problems as the cutting of Dullatur Bog but also tact and determination in handling negotiations with the Carron Company and soothing the numerous quarrels among his employers and subordinates. In his final report, he suggested that 'public notice should be given of the state of the canal', since he had been informed that a shipwright in Perth had built a canal boat 2 ft longer than the locks. By August 1773 the canal had been completed from the Forth to Kirkintilloch, where boats of 50 tons burden were arriving with cargoes from Bo'ness; in addition, the Townhead Reservoir had been filled, and a basin had been laid out at Stockingfield, 2 miles north of Glasgow. By the end of the year, lock-keepers had been appointed for the sea-lock and the locks near Bainsford, their wages being 8s and 6s a week respectively.[18]

In January 1775 the canal was filled as far as Stockingfield to a depth of 5 ft; but in July work on the main line came to a halt because of the lack of funds, the company having now accumulated a debt of £31,000. Further quarrels arose with the Carron Company, first because Mackell closed a temporary cut they had made at Dalderse, and then because of competition for water supplies. Mackell was searching for means of replacing the water which had been taken from the Kelvin, and the Carron Company accused him of diverting streams which had previously supplied their works. The Carron Company having dug a ditch to carry what

Mackell regarded as the canal company's water from the Banton Burn into the Carron, Mackell proceeded, despite threats to 'blow out the brains' of anyone attempting it, to cut the side of this ditch so that the water flowed back into a feeder of the canal. Gascoigne protested vehemently against this 'licentiousness', and denounced the canal company's employees as 'public nuisances'.[19]

In May 1776, because 'present credit was not sufficient' for the continuation of work on the Glasgow branch, where land was valued at around £45 an acre, an urgent request for funds was sent to Sir Lawrence Dundas. Funds must have been obtained, for in October the company was planning a new basin between the Black Quarry and Hamiltonhill, less than a mile north of Glasgow. In January George Chalmers had proposed that the Convention of Royal Burghs should apply for public aid, since it was 'a disgrace to the whole island that the navigation should remain in its present unfinished state'; and in November the Convention wrote to the canal company recommending that a petition be organized. Since the company was in debt to the extent of £40,000 and needed a further £60,000 to complete the original plan, the case for public support seemed strong; and the petition received the backing of all the royal burghs except Glasgow, which insisted that the line to Dalmuir should be abandoned in favour of a continuation of the Glasgow branch to the Broomielaw. Chalmers was sent to London to argue the case; but no assistance was forthcoming.[20]

By November 1777 the Glasgow branch had reached the new basin below Hamiltonhill, which was 'formed of ashlar work' and could hold 'a dozen of vessels of 70 tons burden'; and in February the newly-built *Earl of Dunmore* was being fitted out there, and several sloops were loading and unloading cargoes. By March 1778 £157,327 had been spent on the main canal and £7,169 on the Glasgow branch; and despite statements to the contrary in Priestley's *Historical Account of the Navigable Rivers, Canals and Railways of Great Britain* and subsequent works it seems clear from the records that the canal company had paid for both. Mackell died in November 1779, and because the work was at a standstill no successor was chosen; Nicol Baird, who had been toll-collector at the sea-lock, was appointed surveyor at a salary of £100 per annum, and assumed responsibility for the maintenance of those parts of the canal which had been completed. In March 1780 the Commissioners of Forfeited Estates turned down a proposal that they should lend money to the canal company on the security of the tolls, which had amounted in the previous year to £5,056.[21]

Small cargo-vessels were built to compete with the waggons and

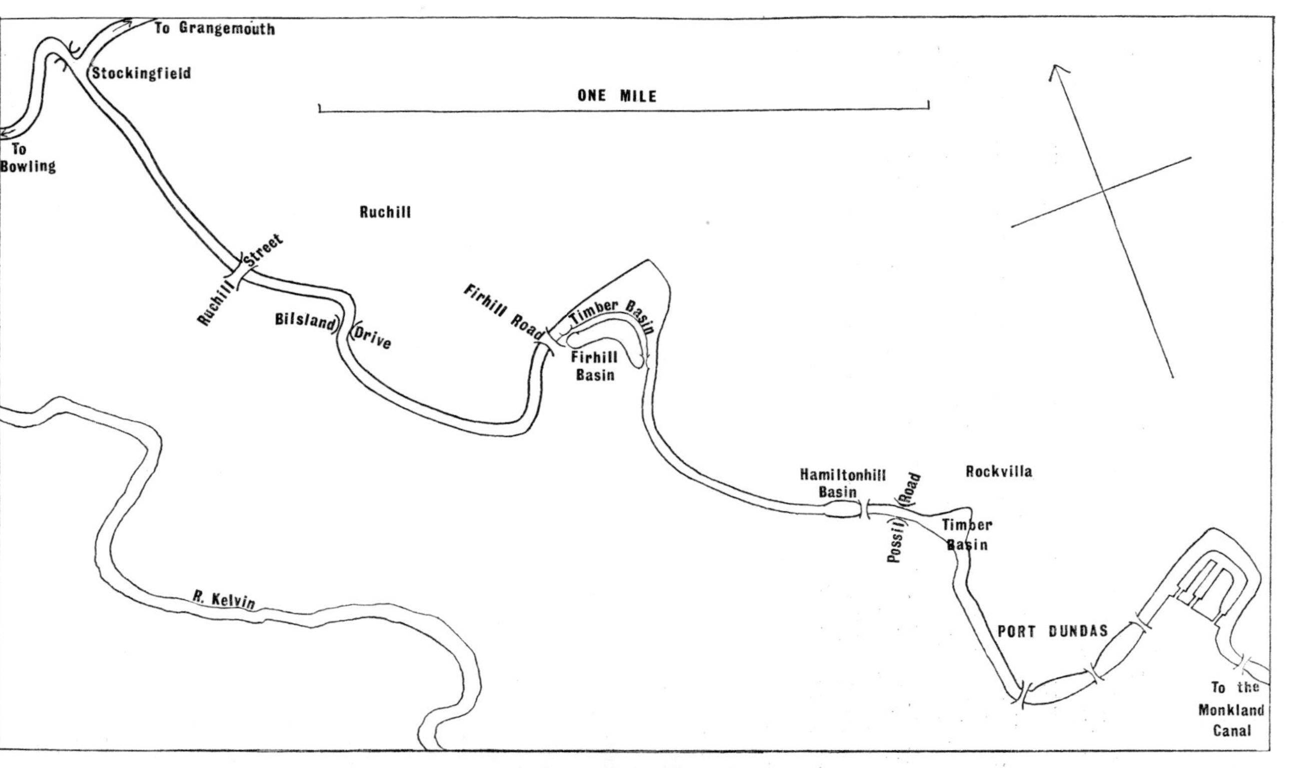

3. The Glasgow Branch

carts operating between Glasgow and Edinburgh; and the canal's value to the community was demonstrated after the bad harvests of 1782 and 1783, when grain shipped from Danzig and elsewhere to Hamiltonhill helped to avert a famine in the west of Scotland. The company's financial difficulties had been aggravated by the American War and the consequent collapse of Glasgow's tobacco trade; but prospects improved with the advent of peace. Sir Thomas Dundas, who had succeeded to the baronetcy in 1781, offered to give the ground and half the money for a navigable cut from the mouth of the Grange Burn to the Forth such as had been suggested by Brindley in 1768: and in May 1783 the company accepted an offer for digging this cut at 3¼d per cu yd, and thus finally abandoned the scheme for an entry beside the Heuck Farm. In the same year, ambitious plans were made for the construction of 'basins, harbours, quays, wharfs and cranes' at the eastern entry, which was soon to be known as Grangemouth.[22]

In May 1784 advertisements were inserted in the Glasgow and Edinburgh newspapers to inform merchants that the track-boats *Glasgow* and *Lady Charlotte* were operating from Hamiltonhill to 'Sea-Lock' and Leith respectively, and that the rates for 'various items' were from 7s to 10s 6d per ton for the shorter journey and from 9s to 14s per ton for the longer one. In March of that year the new Prime Minister, William Pitt, had promised that the company's request for government aid would be considered as soon as possible; and in August an Act was passed directing the Barons of the Exchequer in Scotland to lend the company £50,000 out of the money acquired through the sale of the forfeited estates, and giving the company permission to choose the most convenient point for the canal's entry to the Clyde.[23]

It had been proposed that the whole canal should be made part of the port of Glasgow, so that goods landed at Glasgow and re-exported through it could be deemed to have been re-exported from the point of importation; but this proposal was not incorporated in the Act, and the company was equally unsuccessful in its campaign for the abolition of customs-house fees on goods passing along the canal and of coastal bonds on goods landed within the two firths. The provision of public aid, however, made it possible to begin work on the final stage of the canal's construction; and in December Archibald Millar was engaged to survey the ground from Stockingfield to Dalmuir. Early in 1785 Smeaton was called in to advise on the dam being erected at Grangemouth to divert the Carron into the new navigable cut; it is not clear whether his advice was followed, but the dam collapsed in March, and Baird recommended that the cut should be abandoned. Because of the changes which had taken place in the Clyde, the question of the canal's western entry had to be reconsidered; and in May Millar and Laurie (who had been reinstated) proposed that it should be not at Dalmuir but at Bowling, where new coal-pits had recently been sunk.[24]

Robert Whitworth, who had worked in England under Brindley, became the company's chief engineer in June 1785. In July he reported on the navigable cut at Grangemouth, proposing that it should be deepened and that the upper end should be widened to relieve the pressure on the dam. In August he reported on the proposed continuation of the canal from Stockingfield to the Clyde, agreeing with Millar and Laurie that the western entry should be at Bowling and estimating the cost of the work at £56,456, or £58,901 if the depth were increased to 8 ft; the Kelvin Aqueduct was expected in either case to cost £6,200. In September, reporting on the possibility of increasing the canal's depth by raising the water-level, he discussed the question of water-supplies for this purpose and for the cut to Bowling, and suggested not only that the Townhead Reservoir and the smaller lochs on which the canal already drew should be raised but also that the central part of Dullatur Bog should be made into a reservoir by means of embankments. The Carron Company objected to this proposal, however, and although Whitworth doubted their claims to the water involved, he turned his attention in October to a more ambitious scheme for drawing water from three new reservoirs by way of the Monkland Canal, the locks to be made for the Monkland at Blackhill, and a cut of connection from the Monkland Basin to the end of the Forth & Clyde's Glasgow branch. In January 1786 Millar reported on this scheme in more detail, showing that no locks would be required on the route from the Monkland Basin to Hamiltonhill, and concluding, 'If any part of the canal navigation makes a suitable return, it must be this improvement.'[25]

A pamphlet published in 1786 by an anonymous proprietor listed the canal's works as now including '20 locks, 2 large aqueduct bridges, 10 smaller ones, 15 drawbridges, 14 tunnels, 3 basins for shipping and 2 large reservoirs, 2 ice-boats and 3 track-boats for transporting goods and passengers'. The author complained of the 'exorbitant price' of over £45 per acre which had been paid for poor land, and of such extravagances as Smeaton's salary of £500 for a 'twice-a-year' visit; and he demanded that the company's general meetings be held in Edinburgh, since 1,213 shares were held by natives of Scotland and only 130 by natives of England.[26]

In August Whitworth reported again on water-supplies for the deepening of the canal. The canal could already draw 2,245 lockfuls annually from the Townhead Reservoir, 2,545 lockfuls from the Bishop, Woodend, Gartsherrie and Johnston Lochs, and 300 lockfuls from Possil Loch; and Whitworth was confident that the three reservoirs east of the Monkland would make up the full requirement, so that the dangerous and expensive operation on Dullatur Bog could be abandoned. He estimated the cost of these reservoirs at £4,466, that of the aqueduct which was to take the water from the River Calder to the east end of the Monkland at £561, and that of a canal 4½ ft deep from the Monkland Basin to Hamiltonhill at £2,407. The contract for the Kelvin Aqueduct, a much greater task than the earlier aqueduct over the Luggie at Kirkintilloch, was awarded in November to William Gibb and John Moir of Falkirk. In May 1787 it was enacted that the company should in future be run by a Governor and Council, that all general meetings should be held in London, and that the depth of the canal should be increased to 8 ft. Archibald Spiers, the chairman of the committee, laid the foundation-stone of the Kelvin Aqueduct on 16 June 1787; and in March 1788 James Maxwell published a long poem in the canal's honour, praising the quays and sawmills of Hamiltonhill and giving the following description of the track-boats:

> Lo, here's two vessels built with good design,
> Large and substantial, fitted out right fine
> To carry passengers on this canal,
> Or any other goods when haste may call:
> One to go east, another to go west,
> Which way soever suits the purpose best.
> Here passengers with cheerfulness may go
> On board hereof, above deck or below.
> A pleasant passage they may here enjoy,
> Divest of danger, void of all annoy;
> For here a cabin in each end is found,
> That doth with all conveniences abound;
> One in the head, for ladies nine or ten,
> Another in the stern, for gentlemen,
> With fires and tables, seats to sit at ease,
> That may regale themselves with what they please.

Arrears of £3,538 still included the sums owed by the three Carron Company proprietors who refused to pay until their claims on the canal company had been met; but the canal's revenue totalled £9,764 in 1787, and the company expected a 'steady revenue' of £11,000 after the junction with the Clyde, and an 'immense rise' in the event of war. A new timber basin was laid out at Firhill in 1788: and in April 1789 it was reported that the Kelvin Aqueduct was seven-eighths finished, and had become an 'interesting object of curiosity to all strangers'. The work on the aqueduct had been constantly supervised by Whitworth; and his attention to it was praised by the canal's agent, Patrick Colquhoun, in a letter to one of the shareholders. Later in the year the *Scots Magazine* drew the attention of its readers to the new stretch of canal from Stockingfield to the Kelvin, on which there were '3 aqueduct bridges, 5 locks, and as many circular basins, besides a dry-dock for careening vessels', and declared that the 4 arches of the great aqueduct were 'placed in a situation truly romantic' and would 'form one of the most picturesque scenes' imaginable when the canal was open from sea to sea.[27]

The operations for the deepening of the canal were completed in 1789, and it was expected that they would benefit the country's coastal trade. In the same year conditions were worked out for the junction with the Monkland: the company, at its own expense, was to convey from the Calder to the Monkland a supply of water equal to that which it took from the Monkland, erect a sluice between the upper and lower reaches of the Monkland, and make a new cut linking the two canals; and in return they were to have the free use of the Monkland as a feeder, all repairs except those consequent upon this use being undertaken by the Monkland Canal Company. Plans were made at the same time for a new basin on the cut of communication at 'Hundred-acre-hill', later known as Port Dundas, and for the laying out of 'streets, roads and commodious passages' there to link with the 'streets and lanes of the city of Glasgow'; the basin was to be 900 ft long and 100 ft wide, and there were to be granaries, warehouses, 12 timber yards and a 50-ft wharf. The Act authorizing all these developments was passed in June 1790.[28]

Work continued on the cut to Bowling; and because of the complaints which the Highland drovers had made when their cattle refused to cross the pivot bridges north of Falkirk Whitworth decided that the new bridges should be flat drawbridges with fixed railings along their sides. In November 1789 a boat fitted with a double engine constructed by William Symington of Wanlockhead and Patrick Miller of Dalswinton was tried out on the canal: the experiment was only partially successful, but it was hailed by the *Glasgow Mercury* as being 'of the greatest utility to mankind', and has its place in the history of steam navigation. At the end of the year the men working on the canal included 14 contractors, 25 carpenters, 81 quarry-men, 130 masons and 419 labourers; and it was stated that many other labourers had returned to the Highlands for Christmas. Work had begun on a dry-dock north of the basin at Bowling; and it was expected that the canal would be finished by October 1790.[29]

The official opening actually took place in July, when one of the company's barges sailed from Glasgow to Bowling. The *Scots Magazine* reported:

> The voyage, which is upwards of 12 miles, was performed in less than 4 hours, during which the vessel passed through 19 locks, descending thereby 156 ft from the summit of the canal into the Clyde. It required only 4 minutes to pass each of the locks, in which space the vessel descended 8 ft into the reach of the navigation immediately below.
>
> In the course of the voyage from Glasgow to Bowling Bay, the track-boat passed along that stupendous bridge, the great aqueduct over the Kelvin, 400 ft in length, exhibiting to the spectators in the valley below the singular and new object of a vessel navigating 70 ft over their heads—a feature of this work which gives it a pre-eminence over everything of a similar nature in Europe, and does infinite honour to the professional skill of that able engineer Robert Whitworth Esq, under whose direction the whole of this great work has been completed in a very masterly manner.
>
> The committee of management, accompanied by the magistrates of Glasgow, were the first voyagers upon this new navigation. On the arrival of the vessel at Bowling Bay, and after descending from the last lock into the Clyde, the ceremony of the junction of the Forth and Clyde was performed, in presence of a great crowd of spectators, by Archibald Spiers Esq of Elderslie, chairman of the committee of management, who, with the assistance of the chief engineer, launched a hogshead of water of the river Forth into the Clyde as a symbol of joining the eastern and western seas together.

The canal was 38¾ miles long, about 60 ft wide at water surface and about 30 ft wide at the bottom. The summit, which extended from Stockingfield to near Castlecary, was 156 ft above sea level; and there were 20 locks east of it and 19 locks west of it, each lock being approximately 20 ft wide and 70 ft long. The water for the western part was to come chiefly through the Monkland, and

I. Forth & Clyde Canal: (*above*) the outer basin at Bowling; (*below*) the Kelvin Aqueduct about 1825

II. Forth & Clyde Canal: (*above*) the aqueduct over Maryhill Road; (*below*) the junction with the Glasgow branch, looking south

that for the eastern part chiefly from the Townhead Reservoir. Apart from the Glasgow branch, the only side-cuts were that made by the Carron Company at Dalderse, which had been re-opened in 1782, and that made by Henry Glassford at Netherwood Lime Works; but it was expected that a network of subsidiary canals would soon develop.[30]

The opening of navigation was celebrated in the *Glasgow Courier* by a panegyric to the Kelvin Aqueduct:

> Tho' spiteful Kelvin threatened to divide
> Forth's tumbling flood from joining with the Clyde,
> Thy rising form, Majestic, interpos'd,
> Strode o'er the vale, and the wide gap was clos'd:
> To vanquish Nature's local spite the more,
> The trusty Locks retain their liquid store;
> Which from the height by gradual steps descends,
> Till on thy top the short-liv'd torrent ends.
> How grand the view, when, from the hollow vale,
> The eye delighted sees the coming sail
> With steady pace her middle region ply,
> And, on thy summit, hang 'twixt earth and sky;
> Nor here does end the wonder and amaze,
> Which still must strike the curious stranger's gaze,
> As they their course from West to East explore,
> Or from the East do seek a Western shore . . .
> Thus all the features of this vast design
> In one great cause their mutual efforts join,
> While thy huge Fabric swells above the rest,
> And stands the Monarch of the Group confess'd.

Work continued at Port Dundas, 8 acres around the basin being disposed of in lots. There were plans for a bridge-keeper's house, a collector's house and a customs house; but there was to be no 'noxious or offensive manufacture', and no public house unlicensed by the company. By June 1791 the canal from Hamiltonhill to Port Dundas was finished, and both the main basin and the timber basin east of it were well under way; and in October the cut of junction with the Monkland was opened, and a vessel 'arrived at Hamiltonhill from the eastward amidst the acclamations of a great concourse of people'. The advertisement for a sale of building lots held in September 1792 claimed that the houses to be erected at Port Dundas would each have 'back ground sufficient for a good garden', and that the inhabitants would not only have 'the enjoyment of the country air' but also be 'well-situated for carrying on trade to great advantage with many towns in England, Scotland and Ireland'. The 'many other conveniences' which were to make the village 'a most desirable residence' included the availability of water from the canal and of cheap coal brought by the Monkland. The growing traffic at Port Dundas necessitated the appointment of an assistant bridge-keeper in November 1792; and when the basin was busy it presented a 'very singular' appearance to those passing along the road 70 ft below, since the buildings on the hillside were 'intermixed with trees and masts of vessels'.[31]

Other developments at this time included the raising of the water-level in the Black Loch, the first of the three Monkland reservoirs to be brought into use, and the formulation of a rule whereby the canal was to be closed when ice required the use of more than 10 horses to pull a single boat. The opening of the Blackhill locks in November 1793 completed the through navigation from the Monkland coalpits to Port Dundas, and in January 1794 the company decided to encourage the resultant trade by reducing the toll on coal to ½d per ton per mile, except when water was scarce. The company was still £30,000 in debt, and no dividend had yet been paid; but the revenue rose from £9,764 in 1787 to £12,373 in 1794. One cause of this increase was the large number of west-coast herring-busses passing through the canal to fish in the Forth, where herring appeared 'in great abundance' in 1792. Fifty-four busses passed through the canal in December 1794, all of them being expected to return within a few weeks; and the canal profited again from the appearance of herring in the Forth in September 1795 and March 1796.[32]

In February 1796 several labourers were injured by the collapse of a piling-engine being used to repair the canal bank in Dullatur Bog; and the company therefore decided to subscribe to the Glasgow Royal Infirmary so that employees injured in future accidents would be entitled to receive treatment there. The increase in trade resulting from the junction with the Monkland created an urgent need for new water-supplies; and in July Hugh Baird, the son of the company's surveyor, inspected the proposed reservoir-sites at Hillend Moss and Roughrigg. Work on the former site began in November; and the reservoir there was completed in December 1798. Earlier in that year authority had been given for the erection of a sluice at the top of the Blackhill locks, and for the construction of a tunnel through which water could be conveyed from the sluice into the lower reach of the Monkland. In July 1798 the canal company, which was enjoying an increase in revenue because it could offer a route from east to west which was not threatened by French warships, agreed to 'give dispatch' to vessels carrying troops and military stores and to let them pass free of dues. An Act of 1799 increased the company's stock to £421,525 in 1,297 shares of £325, authorized repayment without interest of the £50,000 government loan, and repealed the restriction on dividends; and in April 1800 the company paid a first dividend of 10 per cent. War-time inflation, however, and the disruption of trade with the Baltic increased the company's labour costs and raised the price of timber for new canal buildings; and in March 1801 it was decided that the dividend should be kept at 10 per cent and that £10,000 should be borrowed from the Royal Bank.[33]

In June 1800 Lord Dundas of Kerse, formerly Sir Thomas, persuaded the company to have a boat constructed 'after Captain Schank's model' and fitted with a steam-engine designed by William Symington. This vessel, which was named the *Charlotte Dundas* after Lord Dundas' daughter, was tried out in a series of experiments between January 1801 and April 1803, and has been described as 'the first practical steamboat'. In March 1803 it towed two loaded vessels from Lock 20 to Port Dundas in 9¼ hours, thus inspiring William Muir of Kirkintilloch to write 'The Steam Barge':

When first by labour Forth an' Clyde
Were taught o'er Scotia's hills to ride
In a canal deep, lang an' wide,
  Naebody thought
That *wonders* without win' or tide
  Wou'd e'er be wrought. . .
But lately we ha'e seen a lighter,
An' in her doup a fanner's flighter,
May bid boat-haulers a' gae dight her
  Black sooty vent;
Than ha'f a dozen horse she's wighter
  By ten per cent. . . .
When first I saw her in a tether
Draw twa sloops after ane anither
Regardless o' the win' an' weather
  Athwart her bearing,
I thought frae hell she had come hither
  A privateering;
An' that the pair she had in tow
Were prizes, struck me sae, I vow,
I cry'd (when fixed to their prow
  I saw her cable),
'In Satan's furnace now they'll lowe
  Amang the rabble'.
It was sae odd to see her pulling,

An' win' an' weather baith unwilling;
Yet deil me care she onward sculling
Defy'd them baith,
As constant as a mill that's fulling
Gude English claith.

The company, however, was worried by the effect of the paddles on the canal banks; and though the possibility of converting the *Charlotte Dundas* into a dredger was considered in 1808, the idea of steam navigation was not taken up again until 1828.[34]

The decline of the fishing industry, which was accelerated by the war and the navy's demand for sailors, reduced the canal's dues on herring from £4,995 in 1800 to £966 in 1803; but other trade increased slightly, and the total revenue was cut by only £314. Under an Act of 1806 the administration of the company was entrusted to a single body consisting of the Governor, four councillors chosen from proprietors resident in London, and three councillors chosen from proprietors resident in Scotland. In the same year there were complaints from coal traders about the shallowness of the cut of junction; and it was agreed that a £1,180 regulating-lock should be built at the east end of the basin at Port Dundas and that the banks of the cut of junction should be raised, provided the Monkland company would raise the banks of their lower reach correspondingly. In February 1808 the company responded to complaints about the unfenced state of the old basin at Hamiltonhill, where a young woman had just been drowned, by ordering Hugh Baird, who had become surveyor on his father's death in the previous year, to erect a railing with 'any old wood' he could find. Michael Johnstone, the master of the track-boat *Lady Catherine*, who had in 1799 saved one of his crew 'at a very great risk of his own life', was drowned in 1809 and lamented by William Muir:

Within this cold grave,
Dragg'd from the dull wave
Of the Forth and Clyde Navigation,
Lies the skipper, ye'll note,
Of the *Catherine* track-boat,
Who with credit did long fill his station.
From Glasgow to Grange
(You may think it was strange)
He could tell the Canal's ev'ry winding,
And each fanciful twist
The whole length of its coast
Was known from beginning to ending.
One would thought that a place

That so well knew his face
To old age his acquaintance would cherish'd;
But alas! at a lock
(With grief be it spoke)
By the side of its sluices he perish'd.
He perish'd, where oft
His bark rode aloft
And shoulder'd the sides of the building,
While free from the cord
And the lash of its lord
Stood in waiting the wet, weary gelding.
Now wanderer here
If thou grudge him a tear
The heart is as hard as a whinstone;
For, his failings apart,
A kind feeling heart
Warm'd the bosom of poor Michael Johnstone.

Despite the general improvement in trade, the canal's financial position was adversely affected by the closing of the Baltic ports and by the need for expensive repairs to locks, bridges and basins; and the dividends for the years 1807–10 remained around 10 per cent. By an Act of 1809 the depth of the Clyde from Glasgow Bridge to Dumbarton Castle was to be increased to a minimum of 9 ft; and in October the canal company brought their dues from Port Dundas to Bowling in line with those on the river by reducing them to 1d per ton per mile on all goods except coal. In February 1812 the company's office was moved from Glasgow to Port Dundas, and Baird was appointed Resident Engineer with a salary of £250 per annum plus £50 to cover 'all contingencies'. In the same year the Carron Company laid out a basin on the north side of the canal at Bainsford; this was linked to the Carron Works by a railway, which replaced the old navigable cut from Dalderse. In August the canal company decided to establish a passenger-service between Glasgow and Edinburgh by way of the sixteenth lock at Camelon. The passage-boat left Port Dundas in the morning, connected with the Edinburgh coach at Camelon at 1 pm, and returned immediately to Port Dundas; and the experiment was so successful that arrangements were made to have four boats in operation the following spring. Dues on grain, sugar, coffee and other goods were raised in February 1813 to cover the rising cost of labour and materials; and in 1814 an Act was obtained for extending Port Dundas, increasing the canal's depth to 10 ft and continuing the towpath to the mouth of the Carron.[35]

Proposals for the improvement of Grangemouth put forward in

1814 sparked off a bitter controversy which reached its climax in 1816 and for a time ended the Dundas family's domination of the canal. Grangemouth, which had not existed half a century earlier, was now a small but flourishing town in the angle formed by the canal and the River Carron. It had a customs house and several 'handsome streets'; and though the war had interrupted its trade with Norway and Sweden it had not only a coasting trade with London but also a considerable trade through the canal in American timber. Besides the inner and outer basins, there was a large timber-basin south-west of the town which could hold '300,000 ft in rafts'; but more space was needed for shipping. Baird, who had submitted plans for extending the harbour in April 1810, put forward in October 1814 a new proposal for a £17,629 wet dock to be entered from the Forth; but preference was given to a more ambitious scheme drawn up by Rennie, which was to cost £125,000. Critics of this decision declared that it was in the interests not of the company but of the Governor, Lord Dundas, who would be able to sell his land at an inflated price. The old controversy about the canal's eastern entry was recalled, stress being laid on the claims of Bo'ness; and it was alleged that the straightening of the Carron had been carried out simply to add '60 acres of fertile land' to the Dundas estates. The opposition, which was headed by the Glasgow cotton-manufacturer Kirkman Finlay, contended also that since the Grangemouth harbour and wharf dues had brought the company only £571 in 1814, there was no reason to suppose that the improvements would pay.[36]

Finlay and others served two bills of suspension and interdict against the company in July 1815 to prevent them from uplifting money from their account in the Royal Bank and using it on the proposed improvements to Grangemouth. The company argued that the value of their stock was attracting speculators in search of short-term profit, and that the improvements were needed to make the navigation 'more commodious'. Allegations of mismanagement were refuted by reference to the steady increase in revenue over the previous thirty years; and it was asserted that the dividend of almost 20 per cent already being paid was the highest practicable until the economic consequences of the peace had become clearer. Dundas' critics, returning to the attack, condemned such 'idle and unprofitable projects' as the steamboat and the new office at Port Dundas, and questioned the practice of holding meetings in London when most of the proprietors resided in Scotland. In March 1816 Kirkman Finlay was elected Governor, the resolutions authorizing the new works at Grangemouth were rescinded, and the dividend was raised to 25 per cent. This ended Dundas' thirty-year reign as Governor, during which the company's revenue had risen from £6,835 to £46,974; he died in 1820.[37]

Since the 1790s the company had had four track-boats which served mainly for carrying goods but also had limited cabin accommodation for passengers; but it was not until August 1806 that they considered the possibility of having two boats fitted up 'in complete manner for passengers alone'. The proposal was brought forward again in February 1808; and on 30 May 1809 two boats began to operate between Port Dundas and Lock 16. The number of passengers carried in 1808 was 6,771, and the total revenue from passenger traffic was £580; by 1815, when three boats were making double trips, these figures had risen to 85,368 and £7,087. The boats were 66 ft long and 10½ ft wide, and drew 2½ to 3 ft of water. Each of them had 'a cabin, a steerage, a small eating-room and a place for stores'. The cabins were 'elegantly fitted up', and the passengers were 'accommodated with newspapers, books, backgammon-tables, etc'. Breakfast and dinner were available 'at moderate rates', and 'wine, strong ale and porter' were on sale. The journey from Port Dundas to Lock 16 involved passing through 4 locks and took 5½ hours; and the fares were 4s cabin and 2s steerage. Between Camelon and Grangemouth, where there were 16 locks in 3 miles, there was only one boat per day in each direction; but in May 1816 the company altered their boats' times of departure from Port Dundas to 7 am, 11 am and 5 pm so that passengers could proceed from Camelon to Grangemouth by coach and from there to Leith by a 'steam vessel tug' of the Edinburgh, Glasgow & Leith Shipping Company, the coach returning to Camelon with the passengers the tug had brought from Leith. Newspaper advertisements declared that a 'comfortable and cheap conveyance' between Scotland's two chief cities had thus been established.[38]

In March 1818 the company considered plans for a new harbour and wet dock at Bowling, and made arrangements with Lord Dundas for the construction of a new timber-basin west of the existing one at Grangemouth. In 1819 a 'malleable iron passage-vessel' named the *Vulcan* came into operation on the canal: constructed of iron from the Monkland Steel Works by Thomas Wilson of Faskine to the plans of Henry Creighton, it was 63 ft long, 13 ft wide and 5 ft deep, had 'a large cabin and awning on deck' and could carry 200 passengers and their luggage. In the same year proposals were made for extending the canal from Bowling towards Dumbarton and for constructing a branch from Lock 20 to meet the still-unfinished Union Canal above Camelon and thus reduce the number of locks on the journey between Glasgow and Edinburgh; but these were never carried out. An Act of 1820 increased the company's stock to £519,840. Passage-boats travelling between Port Dundas and Lock 16 were in this period supplied with new horses at the Glasgow Road Bridge west of Kirkintilloch, and at Craigmarloch where the feeder from the Townhead Reservoir joined the canal; the route was thus divided into three stages of 8 to 8½ miles in length. A long frost early in 1820 reduced the company's revenue for the first quarter of that year by £2,000; and the annual dividend was reduced to 20 per cent. A desire for economy made the company keep a secret watch in May 1821 on six labourers who were repairing the banks between Craigmarloch and Lock 16: it was reported that they were guilty of 'coming late in the morning, quitting their work during the day, absenting themselves from work, and idling their time whilst on the banks', and they were dismissed and evicted from their company houses. In the following month the company agreed to transport free of dues stones to be used in the Burns Monument at Ayr.[39]

In 1822, when the Union Canal was opened, the company withdrew its support from the Camelon–Grangemouth coaches, assuming that passengers would now travel between Camelon and Edinburgh by the Union passage-boats. By August, two new boats were ready to carry goods between Port Dundas and the Union Canal's Edinburgh basin of Port Hopetoun. Extra boats and horses were provided in that month to carry sightseers going to Edinburgh for George IV's visit; and it was decided that they should be permitted to operate on Sunday to 'prevent disappointment to the public'. In December 1823 the company gave its support to the bill for the Monkland & Kirkintilloch Railway, hoping to benefit from the transportation of Monkland coal to Edinburgh by way of Kirkintilloch and Camelon. Because of drastic reductions in the stage-coach fares, however, the company was forced to reduce its passenger rates and revise its timetable for the boats operating between Port Dundas and Lock 16. The new fares were 3s cabin and 2s steerage; and the number of stages was increased from three to six, horses being changed at Cadder, Kirkintilloch, Shirva, Auchinstarrie and Castlecary. Charges for refreshments on board were also reduced; breakfast was now offered for 1s, cold dinner for 1s, hot dinner for 1s 6d, tea or coffee for 1s, and port or white wine for 5s per bottle.[40]

In February 1824 the London, Leith, Edinburgh & Glasgow Shipping Company was given permission to run a night service of luggage-boats carrying steerage passengers between Port Dundas and Port Hopetoun: these became known locally as 'hoolets', presumably because their whistles suggested the hooting of owls. Later in the year another iron boat was made by Thomas Wilson at Tophill; and in September the company ordered two iron mineral barges to handle the new coal trade to be produced by the opening of the Monkland & Kirkintilloch Railway, from which the company expected 'permanent increases' in revenue. The Clyde Navigation Bill of 1825 was opposed by the company because of its effect on the dues paid by vessels entering and leaving the canal at Bowling. In February 1826 the company agreed to subscribe to the proposed Ballochney Railway, a branch of the Monkland & Kirkintilloch, because 'the state of the money market in Glasgow' was making it difficult to get funds. The depression of 1826–7 created considerable unemployment in Glasgow; and the company took advantage of the ready availability of labour to insist on instant dismissal for any employees who refused to give up 'the habit of intoxicating themselves occasionally'. The masters of passage-boats were instructed at the same time to be 'polite and obliging to every passenger without exception'.[41]

In July 1828 the company considered a proposal for the introduction of steamboats, and in October the Inspector of Works pointed out that 'dispatch' was 'everywhere the order of the day in conveying either goods or passengers', and suggested that the banks might be lined with stone so that higher speeds would be practicable either with steam or with horses. It was alleged that the masters of the passage-boats had hitherto behaved 'like innkeepers who wanted to detain passengers as long as possible'. A steamboat named the *Cupid* was tried out on the canal, and achieved an average speed of 3 mph and a maximum speed of 6 mph with a loaded scow in tow; and in January 1830, after Thomas Grahame had deplored the 'distrust and unwillingness' with which some of his fellow-proprietors approached the idea of steam navigation, it was agreed that John Neilson should be asked to adapt a track-boat for steam power in accordance with plans Grahame had procured from New Orleans. The reconstructed boat, which was named the *Cyclops* and described as 'the first steamer built for the canal carrying-trade', made a successful first voyage from Grangemouth to Alloa and back to Port Dundas in the autumn; but the damage to the banks, as in all these early experiments with paddle-steamers, was considerable. In the meantime the company had been following the lead of the Paisley Canal in experimenting with lighter and swifter boats drawn by horses. A 'twin gigboat' constructed by lashing two ordinary ones together had had a successful trial in April; and on 7 and 8 July a twin boat named the *Swift*, 60 ft long and fitted to carry 50–60 passengers, had made the journey from Port Dundas to Port Hopetoun in 7 hours 14 minutes and the return journey in 6 hours 38 minutes. Orders were therefore given for the construction of a twin steamboat 68 ft long; and the task was entrusted to William Fairbairn. The chief competition in the 1830s came from faster coaches running on the improved turnpike roads; but in March 1831 Grahame returned from a visit to Manchester, where he had seen the trial of this boat, the *Lord Dundas*, and reported that there was a 'universal outcry in favour of railways' and that their promoters were confident 'that in time they must supplant all canal communication'. He believed that the use of steam power would enable the canal to beat the railways; but he noted that the prospect of railway competition had already brought the value of Forth & Clyde shares down from £650 to £570. Another steamer, the *Manchester*, was brought into service between Port Dundas and Alloa in 1832.[42]

The Forth & Clyde was fortunate in that its trade came 'chiefly from the direct passage of sea vessels through the canal or between the city of Glasgow and a number of British and foreign seaports', so that it was less exposed to competition than other canal undertakings. Some of the proprietors were in any case contemptuous of railway competition, and expected the Edinburgh & Glasgow Railway, proposed in 1832, to prove ruinous to its promoters. The new threat did, however, prompt a number of attempts to improve the company's passenger and goods services. The *Lord Dundas* was expected to maintain a speed of 6 mph; and an experimental horse-drawn boat tried out in 1831 did the journey from Port Dundas to Lock 16 in just over 3 hours. New passage-boats modelled on those of the Paisley Canal were brought into operation in 1833; and at the same time new coach-services were established from convenient points on the canal to Stirling and places beyond. Stables were erected at Stirling for the company's horses in August 1835; and the total time for the journey from Port Dundas to Stirling was given as 3¾ hours. One traveller who went from Perth to Port Dundas by coach and passage-boat recorded his pleasure in transferring 'from a rumbling old drag, badly horsed and worse driven, to a snug and warm cabin in the Edinburgh and Glasgow barge, which went at the rate of 9½ miles per hour throughout the whole journey'. Thanks to these improvements, the number of passengers carried by the company's passage-boats rose to 197,710 in 1836. Changes were also made in the goods service between Glasgow and Edinburgh, the time for the journey being reduced from about 15 to about 10 hours by the introduction of lighter boats. Other experimental vessels included an iron 'cart-boat' put into service in 1833, which carried from 16 to 18 carts, and a number of 'waggon-boats' introduced in 1835. Each of these could carry 14 waggons with a total load of 40 tons of coal, and the waggons could run directly on to the deck from the end of the Monkland & Kirkintilloch Railway; the experiment proved highly successful. The canal's effect on the community, however, was not consistently beneficial. In 1832 the passage-boats had to be stopped for a time because the Hamburg-centred cholera epidemic which was affecting several places on the east coast had also spread to Kirkintilloch, Maryhill and Glasgow; and in 1836 the Rev Andrew Sym of Kilpatrick complained that the funeral expenses of people drowned in the canal were seriously reducing his parish funds.[43]

In May 1836 authority was given for the construction of a half-mile canal from the Forth & Clyde at Whitecrook to the Clyde opposite the mouth of the Cart, which was navigable as far as Paisley. This cut was completed in 1840 and taken over by the Forth & Clyde in 1855; but it was never a commercial success, and was closed in 1893. A new reservoir at the Lily Loch was completed later in 1836 at a cost of £914; and in the same year a new basin costing £767 was opened at Port Dundas. In 1837 Grangemouth became a bonded port for all goods except tobacco, East Indian products and European silk; and in 1838 agreement was reached with Dundas' son, the Earl of Zetland, for the diversion of the Grange Burn and the construction of a wet dock entered from the Carron. In the same year the completion of the Newcastle & Carlisle Railway brought competition for coast-to-coast traffic.[44]

In August 1839 the company experimented with the use of locomotives to pull its boats: the experiment, which was carried out on a half-mile stretch of railway laid near Lock 16, was successful; but it was decided that the cost of extending such a railway would be too high. The opening of the Slamannan Railway, which connected with the Union Canal to establish a new route from Glasgow to Edinburgh, reduced the passage-boat revenue considerably, the number of passengers carried falling from 2,284 in October 1839 to 978 in October 1840. A handbill of 1841, however, shows that passengers could still leave Port Dundas at 7 am, 9 am or 12 noon for Falkirk, Stirling and Edinburgh, at 9 am for Perth and Kirkcaldy, and at 4.30 pm for Alloa; fares ranged from

3s and 2s for the journey to Falkirk to 12s and 8s 6d for that to Perth. An Act of 1841 authorized extensions to Port Dundas and Grangemouth, which were to be financed by increasing the company's capital stock to £648,500. In the following year the cut of junction was deepened at a cost of £11,000, and authority was obtained for the extension of the Hillend Reservoir and the creation of new reservoirs at Roughrigg and Bogfoot.[45]

The opening of the Edinburgh & Glasgow Railway in 1842 brought a period of fierce competition. In October the company reduced its dues on grain to ½d per ton per mile; and in February 1843 dues on all goods passing between Glasgow and Edinburgh were reduced to the same figure. The company saw little prospect of recovering the lost trade, however, being convinced that the railway company's object was 'less a present return than to run off the canal carriers and secure to themselves a monopoly'. In July it was decided that the passage-boat fares between Glasgow and Edinburgh should be fixed at 3s cabin and 2s steerage; the third-class railway fare was said to be 4s, and there was also a fourth class. Because of the increase in the company's stock, the dividend for 1844 was only 4 per cent, though the tonnage on the canal and the number of passengers carried were actually increasing. In February 1845 the company resolved to make the cabin fares twice those for steerage passengers, in order to 'prevent steerage passengers taking the cabin'. An Act passed in May gave the company the right to act as 'common carriers' and to carry for 'their own profit upon their own navigation animals, goods, wares and merchandise'. This proposal had been resisted by the merchants and traders of Glasgow, who maintained that the opening of the Edinburgh & Glasgow Railway had 'destroyed an injurious monopoly' and created a healthy measure of competition for trade between the cities, but that the canal company would now be able to drive the private operators out of business and come to terms with the railway against the public interest. An Act of July 1845 gave the company the right to make a cut from Lock 20 to near the Union Canal's third lock; but it was never made. The possibility of an amalgamation with the Monkland was discussed in the following year, and it was agreed that this was desirable, provided it was recognized that three shares in the Monkland were equivalent to one share in the Forth & Clyde. This condition was approved by arbiters in March, and the necessary Act was passed in July. A more ambitious amalgamation scheme was announced by the directors of the Edinburgh & Glasgow Railway in February 1846. Under this scheme, the railway company and the Forth &

Clyde were to unite in purchasing the Union Canal, guaranteeing its proprietors an annual dividend of 50s per share. The terms proposed were highly satisfactory to the Forth & Clyde; but in October the railway company changed its mind, and the plan was dropped.[46]

An Act passed in August 1846 authorized the extension of the canal's terminal basin at Bowling and the construction of an additional lock and an outer harbour at a total cost of £21,000. In the following year a new timber-basin was planned for Port Dundas, and eight lighters were ordered from William Napier at £1,130 each. The Act for the amalgamation of the Edinburgh & Glasgow Railway and the Union Canal was passed in 1849, the Forth & Clyde offering no opposition; and in the same year the Drumpellier Railway, which had been designed as a 'feeder' for the Monkland, was bought for £12,000. The improvements at Bowling being completed, work went ahead with the new timber-basin, for which a site had now been found at Firhill. An employees' sick fund had been formed in 1848, and its first report showed a membership of 369 and gave the allowance on a member's death as £6 10s.[47]

In the later 1840s dividends were gradually raised again, reaching a level of 6 per cent in 1849. Trade increased at the western end of the canal, the numerous works between Maryhill and Blackhill being supplied with flax, tallow, grain, timber, ironstone, coal, sugar and other goods brought from Bowling. It was reported in 1850 that the number of vessels using the canal had risen in eight years from 5,496 to 10,785; and in 1851 the company decided to appoint a keeper for each lock. Tonnage was also increasing at Grangemouth, and it was resolved that dues on goods loaded and unloaded at the wet dock there should be reduced to 1s per ton. A new agreement with the Edinburgh & Glasgow and Caledonian Railways was concluded in 1852, the canal giving up its carrying-powers in return for £5,500 annual compensation; and the Governor, the Earl of Zetland, claimed that the 'severe competition' with the railways was now at an end. The Caledonian withdrew from the agreement in 1854; but the company managed to obtain similar terms from the Scottish Central Railway and the Edinburgh, Perth & Dundee Railway.[48]

The Crimean War had no significant effect on the company's trade, but the hope was expressed that trade with northern Europe might improve when it was over. The increasing number of steamers frequenting the Grangemouth harbour made it desirable to extend the accommodation there, and it was decided that this could best be done by enlarging the Junction Canal between the timber-basin and the wet dock. More houses were also needed, and plans for these were sent to the Earl of Zetland in 1855. Existing agreements with the railway companies came to an end in 1857; but new arrangements were made with the Caledonian and the Edinburgh & Glasgow, fixed rates being established in the Edinburgh and Glasgow areas. A dividend of 6 per cent was paid in 1858; and in the same year the company started a fund for widows and retired employees: widows were to be paid £2, and retired employees a pension of 6d per day 'during the pleasure of the board'. In January 1859 the canal company informed the railway companies that it intended to make a branch railway from Grahamston to Grangemouth. Tonnage rates were discussed by the joint committee of the railway and canal companies in 1859, the railway representatives proposing that the minimum rate should be reduced from 8s to 6s 8d because the higher rate had resulted in a diversion of traffic to the canal. The opening of the Helensburgh Railway in this year reduced the traffic on the canal's western end, but revenue increased again in 1860, and a dividend of 6¼ per cent was announced. Work on the Junction Canal was completed in October.[49]

In the years 1856–66 steps were taken to replace horse power on the canal by screw-propulsion. The *Thomas*, a goods lighter with a screw propeller, was brought into operation in September 1856; its speed was 4½ to 5 mph, and its engines and boiler cost £320. By 1858, another 5 lighters had been fitted with screw propellers; and by 1859 there were 18 such vessels in operation and at least 9 more being made ready. One of the boats introduced in this year was an ice-breaker, which was also used as a crane-boat; the engines and boiler for this cost £430. Engines costing £150 were then fitted in the mineral scows which had hitherto been worked by 2 boatmen, 1 horseman and 1 horse apiece; and the number of screw vessels working on the canal rose to 25 in 1860, 36 in 1862, 50 in 1864, and 70 in 1866. By April 1860 the Edinburgh & Glasgow Railway Company were operating the branch line to Grangemouth; and in the following year a 'good understanding' with the railway companies was still said to exist. The half-yearly report for April 1864 reported 'continued prosperity' and recommended a dividend of 7 per cent; and plans were made for a new timber-basin to be entered from the wet dock at Grangemouth. In April 1865 the Earl of Zetland reported that revenue had fallen because of the severe winter and the low grain imports from the Baltic, and recommended a dividend of 6½ per cent. In the years 1866–7 the company's revenue was adversely affected by a strike in the Monkland collieries; but the main focus of interest in these years was the progress of negotiations with the railway companies. These culminated in 1867, despite strong opposition from the North British Railway Company, in 'An Act for vesting in the Caledonian Railway Company the Undertaking of the Company of Proprietors of the Forth & Clyde Navigation'. The undertaking included the Monkland Canal, the Forth & Cart Canal, the Drumpellier Railway and the Grangemouth Branch Railway; but the Caledonian's principal object in the transaction was to gain control of the docks at Grangemouth. The canal's proprietors were to receive an annuity of £71,333, which was 6¼ per cent of the company's capital; and the interests of the North British Railway Company were safeguarded by clauses fixing uniform dues at Grangemouth docks and giving them the right to use the railway from Grahamston to Grangemouth and the basins and quays at Bowling, Kirkintilloch and Dundyvan.[50]

The shareholders of the canal company went on receiving their 6¼ per cent dividend until 1881, when they were given £156 5s of Caledonian Guaranteed 4 per cent Annuities Stock for every £100 of canal stock. All towing on the canal continued to be done by horses, no tugs or steamers being employed for this purpose; but by 1906 60 to 70 per cent of the traffic was being conveyed by steam lighters, the canal being regularly used by 160 such lighters and by about 170 other vessels, most of them scows, barges or sailing-boats. The goods carried in the later nineteenth century included grain, salt, sugar, oil, stones, slates and flour; but the major items were coal and pig-iron, and the only items in which there was an increase between 1867 and 1906 were pig-iron and timber. Tonnage fell from 3,022,583 in 1868 to 955,050 in 1906; and revenue dropped in the same period from £87,145 to £40,108. The Caledonian attributed this decline to the public preference of 'railway connection' to 'canal connection', to the abandonment of some works on the canal and the tendency for new factories to be built near railways, to the fact that Glasgow was turning from the Baltic to America for its grain supplies, to the introduction in the coasting trade and the trade to the islands of steamers too large for the canal's locks, and to the establishment of a line of coastal steamers linking Liverpool with east coast ports from Newcastle to Aberdeen.[51]

The structure of the Kelvin Aqueduct and of other aqueducts over railways, main roads and city streets made it impossible to increase the canal's depth beyond 9½ ft. The railway company claimed, however, that they were keeping the canal in a 'high state of efficiency', and making it a 'better canal' than ever before. Stoppages through frost were brought to an end by the use of ice-breakers pulled by fifteen to twenty horses and pushed by steam-boats; and two steam dredgers were brought into operation. It was said that maintenance costs were increasing because of the wear and tear on the banks caused by the greater use of steam power; but in fact the amount spent on maintaining the canal dropped from £12,190 in 1884 to £11,601 in 1906. Considerable improvements took place in the docks at Grangemouth; but these had by this period become the docks rather of an international port than of a canal terminal. The 1909 report of the Royal Commission on Canals and Waterways described the Forth & Clyde as 'the most important waterway in Scotland' in volume of trade; but it pointed out that the canal's isolation and its small locks placed it at a disadvantage in competing with the railways.[52]

The advent of cheap railway travel did not entirely destroy the canal's passenger traffic. The private company to which this traffic had been leased in 1852 introduced a screw steamer, the *Rockvilla Castle*, about 1860, and continued to operate it until about 1883; and in 1893 the first of the 'Queen' steamers was brought into service by James Aitken & Company of Kirkintilloch, in order to open up 'the unknown reaches of the canal' as a pleasure route. This boat, the *Fairy Queen*, had accommodation for 200 passengers, made the return journey between Kirkintilloch and Port Dundas twice daily, and operated evening cruises from Kirkintilloch to Craigmarloch and back to Port Dundas. It was succeeded in 1897 by a larger vessel, *Fairy Queen II*; and the cruises proved so popular that Aitken's introduced a second boat, the *May Queen*, in 1903 and a third, the *Gipsy Queen*, in 1905. *Fairy Queen II* was disposed of in 1912, and the *May Queen* in 1918; but the *Gipsy Queen*, the largest steamer ever to operate on the canal, continued to offer 'popular pleasure sails across Bonnie Scotland' from Port Dundas to Craigmarloch until the outbreak of the Second World War. The affection in which it was held is reflected in some verses reprinted in 1939 in a leaflet advertising canal trips as 'a delightful change from road travel':

> There's craft upon the river broad,
> There's craft upon the sea,
> And inland on our bonnie lochs
> Still other craft there be;
> But there's a little boat I know
> Whose charms outweigh them all—

III. Forth & Clyde Canal: (*above*) stables near Kirkintilloch; (*below*) the Kirkintilloch wharf and boatyard

IV. Forth & Clyde Canal: (*above*) the *May Queen* and *Fairy Queen II* at Craigmarloch about 1904; (*below*) Lock 16, showing the Union and Canal Inns and the site of Port Downie

She sails (to use the Doric) on
  The Forth and Clyde 'Canaul'!
From Port-Dundas to 'Caurnie' town
  She winds her wooded way,
And on to calm Craigmarloch, there
  A little while to stay;
And when she blows her siren shrill
  To give the bridge-men call,
She fairly wakes the echoes on
  The Forth and Clyde 'Canaul'!
Oh, river boats are bonnie boats,
  And deep-sea boats are grand,
And dear to me's the little boat
  That skirts the 'Silver Strand':
But, ah! the boat I really love—
  That holds my heart in thrall—
I glint her thro' the trees that fringe
  The Forth and Clyde 'Canaul'!

Goods traffic on the canal remained important until the outbreak of the First World War: Salvesen's, the Rankine Line and the Leith, Hull & Hamburg Company each had a fleet of canal boats plying between Glasgow and Grangemouth. The war-time closure of Grangemouth docks to merchant shipping, however, had a disastrous effect on this trade; and the resumption of business after the war was hampered by the competition of steamers sailing between the Clyde and the continent, by the reduced fuel consumption of the Glasgow Corporation Phoenix Electrical Works at Port Dundas, and by the depressed state of the iron and steel industry in particular and of post-war trade in general.[53]

In January 1920 a 'leading traffic authority' suggested that in view of 'the trend of modern transport' it would be expedient to drain the canal and build a high-grade road along its bed. In the 1920s there were arguments about the replacement of the canal's drawbridges with swing bridges operated by electricity. The drawbridges at Bainsford and Camelon had been replaced with swing bridges in 1901, and in 1923 the Caledonian Railway Company made an unsuccessful attempt to get the Falkirk & District Tramways Company, for whose benefit the new bridges had been built, to pay the wages of the five additional men required to work them. In 1926 the county councils of Stirling and Lanarkshire promoted the Forth & Clyde Navigation (Castlecary and Kirkintilloch Road Bridges) Provisional Order, under which they were entitled to replace two of the canal's drawbridges with swing bridges to be maintained and worked at their expense. Traffic on

D

the canal declined from 697,220 tons in 1913 to 136,182 tons in 1930 and 27,571 tons in 1942; and the Clyde Valley Regional Plan of 1946 recommended that the canal should be abandoned and filled in, since it was a serious obstacle to the improvement of road and rail communication, especially in the Glasgow area.[54]

The Caledonian Railway Company had been incorporated in 1923 in the London, Midland & Scottish; and in 1948 the Forth & Clyde was taken over, like the other still-functioning canals, by the British Transport Commission. Pressure for the closing of the canal came from the Burgh of Clydebank and from some other local authorities; and in the 1950s there was a public outcry about the number of drownings, which was given by Glasgow Corporation as sixty in five years. A British Transport Commission report of 1955 declared the canal to be of 'considerable importance' as a supplier of water to industry but of 'negligible value' as a transport undertaking. Its principal traffic was now in oil, and the chief prospect of future traffic was in the transporting of oil and other liquids from the oil refinery and chemical factories at Grangemouth to Glasgow and Clydeside; if this trade did not develop, there would be no serious objection to the closure. In July 1957 the Minister of Transport told the MP for Glasgow (Maryhill) that the canal had been used in 1955 by 171 pleasure-boats, 119 fishing-boats and 15 cargo-vessels, and in 1956 by 139 pleasure-boats, 98 fishing-boats and 14 cargo-vessels.[55]

The 1958 report of the Committee of Inquiry into Inland Waterways stated that the Forth & Clyde's terminal basins at Bowling and Grangemouth were still being used by cargo-vessels, but that no goods were now being conveyed by the canal. The collapse of the canal's carrying-trade was attributed to the smallness of the locks, the lack of facilities in the canal-side industrial plants for receiving or dispatching goods by water, and the fact that the canal was unable to compete with the road and rail transport which industries were now organized to use. The canal was still being used by the small herring-boats of the two firths; but the number of passages made by such boats had gone down from 189 in 1953 to 98 in 1956. The canal remained important as a source of water-supply, but it had no future as a commercial navigation; and its closure would save local authorities considerable sums of money, both on the maintenance of existing bridges and on the construction of new ones.[56]

The proposal that the canal should be closed provoked vehement protests from representatives of the fishing communities on the Firth of Forth; but the government was forced to take a final decision by the progress of the Denny bypass on the Glasgow to Stirling road, for which a £160,000 lifting bridge would have been needed in 1963 if navigation rights on the canal had not been extinguished before that date. The decision to close the canal was announced in June 1961, and gave rise to nostalgic reminiscences in the columns of the *Glasgow Herald* about the steamers which had operated between Port Dundas and Craigmarloch. The Forth & Clyde Canal (Extinguishment of Rights of Navigation) Act was passed in March 1962; and the canal was closed on 1 January 1963. In 1965 most of the waterway remained, and one could still visit the Townhead Reservoir and the wharfs of Hamiltonhill and Port Dundas, and walk by the towpath from the basins at Bowling over the Kelvin Aqueduct and past a series of converted or derelict stables and warehouses to the old basin and harbour at Grangemouth. Since then, some sections have been filled in, and many of the opening bridges have been replaced either by flat bridges or by culverts; but there remains, in the Maryhill district of Glasgow, a group of basins, locks and aqueducts which deserves to be maintained as an industrial monument.[57]

CHAPTER II

The Monkland Canal

THE growth of Glasgow's population in the eighteenth century combined with the expansion of trade and industry to create an increased demand for coal; and there were many complaints about 'the increasing price and the scarcity of coals in the city of Glasgow at some seasons'. The local coalmasters were accused of deliberately keeping the prices high; and the heavy cost of land carriage made it uneconomical to bring supplies from pits at any distance from the city. In 1769, therefore, the magistrates of Glasgow discussed plans for 'the opening of communication with some distant collieries by water'; and the Monkland collieries in north Lanarkshire were chosen as those which 'bade the fairest to reduce the price of coal'. James Watt was invited to plan a canal from these collieries to the city; and on 12 December 1769 he wrote to Dr William Small, partner of Matthew Boulton:

> Among other things I have projected a canal to bring coals to the town; for though coal is everywhere hereabout in plenty, and the very town stands upon it, yet measures have been taken by industrious people to monopolise it and raise its price 50 per cent within these 10 years. Now this canal is 9 miles long, goes to a country full of level free coals of good quality, in the hands of many proprietors, who sell them at present at 6d per cart of 7 cwt at the pit.

In his report, Watt put forward two schemes. Each of them was for a canal 32 ft wide at surface, 16 ft wide at bottom and 3 ft deep, with a summit 246 ft above sea level, beginning from Woodhall and running by the banks of the Calder through the collieries and past 'Cotesbridge' to Drumpellier, with a branch to Gartsherrie; but whereas the first scheme, to take the canal from Drumpellier via Camlachie to the Clyde near Glasgow Green, required 25 locks and cost £20,317, the second was to be lockless and finish near Germiston, the terminus being linked with Glasgow by a waggon-way and the total cost being only £9,653.[1]

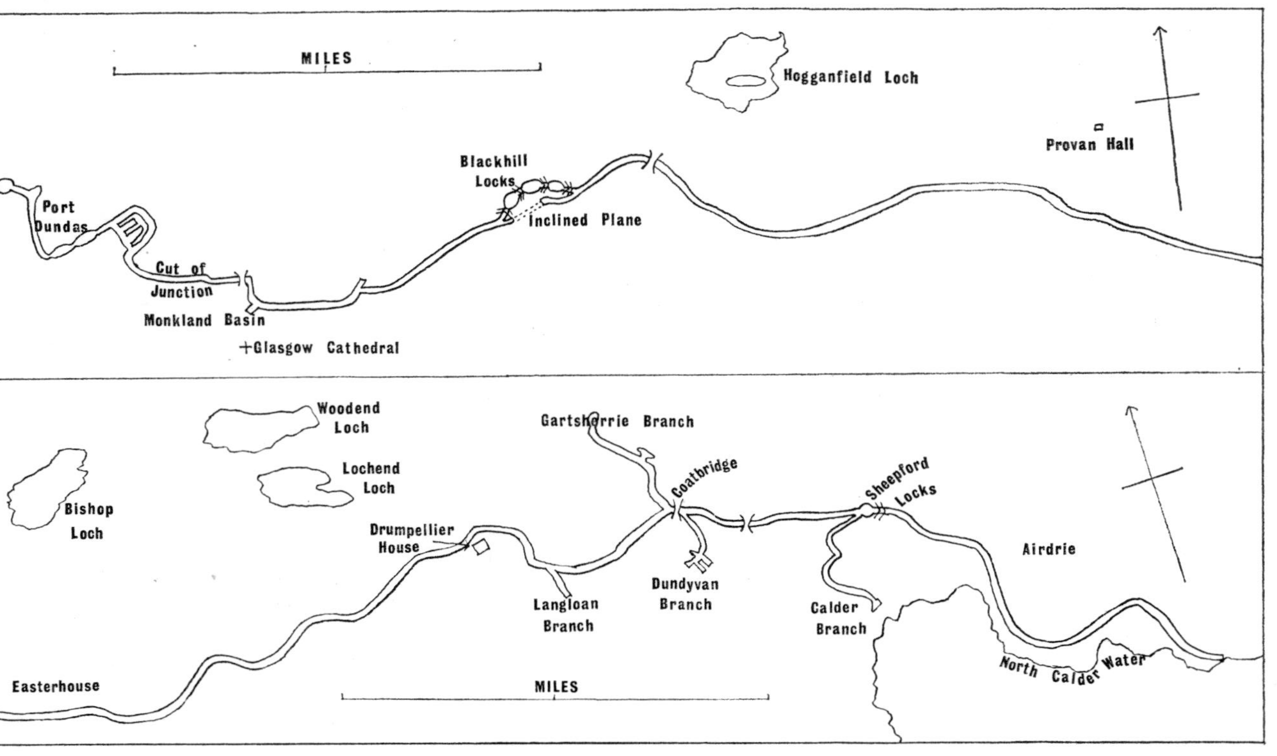

4. Monkland Canal

At a public meeting on 3 January 1770 the organizing committee declared their support for the second scheme, and a subscription list was opened. The town council of Glasgow at first made their subscription conditional on the coal-owners along the route being bound to maintain an annual output of 30,000 tons for 30 years after the canal's completion, but were persuaded to drop this condition in favour of a stipulation that at least £5,000 should be subscribed by proprietors of the land through which the canal was to pass. This demand was easily satisfied, the subscribers including James Dougal of Easterhouse, James Buchanan of Drumpellier, Robert Dick of Gartsherrie, Archibald Hamilton of Monkland and Daniel Campbell of Woodhall. Other subscribers included prominent tobacco-merchants like Alexander Spiers of Elderslie, John Glassford of Dougalston and James Ritchie of Craigton, public bodies like the Trades House and the University, and trades corporations like the maltmen, the wrights, the bakers, the masons and the fleshers—though these organizations subscribed too late to be named in the Act, and had to be assumed as proprietors by the original subscribers at a general meeting of 1 August 1770. The University's contribution aroused the wrath of one professor, who declared that 'if it was for the advantage of the College to have cheap coals, so would it be to have cheap shoes and cheap shirts, and thus the College money might be risked in a tan-work or a linen manufactory'. Among the merchants who subscribed was William Stirling, whose sons were later to be the canal's sole owners.[2]

The Act was passed on 12 April 1770. The preamble claimed that the proposed canal would 'tend to the improvement of the adjacent lands, the relief of the poor and the preservation of the public roads', but laid more stress on the fact that it would 'by reducing the price of pit coal be of great advantage to the trade and manufactures of the city of Glasgow'. The eastern terminus was to be not at Woodhall but in the heart of the collieries at Sheepford; but the route from there to Germiston was to be that suggested in the latter part of Watt's report. Water-supplies were to come from the Frankfield and Hogganfield Lochs and from any streams and lochs within 3 miles of the route which had not already been appropriated for the Forth & Clyde. The proprietors were authorized to raise £10,000 among themselves in shares of £100, and if necessary to raise a new subscription of not more than £5,000. Tolls were to be 1d per ton per mile on coal, stones, timber and dung fuel, ½d per ton per mile on ironstone, and ¼d per ton per mile on lime; paving-stones and manure were to be carried free.[3]

Watt agreed to supervise the construction, and workmen began digging at Sheepford on 26 June 1770. The route was divided into lots of 110 yd each, and these were auctioned to contractors. The first lot went to James Johnston at 1¾d per cu yd; but as the weather was bad and the soil was heavy clay Watt estimated that the work cost Johnston 3d per cu yd in wages alone. Some of the contractors proved very troublesome, and on 9 September 1770 Watt wrote to Dr Small: 'Nothing is more contrary to my disposition than bustling and bargaining with mankind; yet that is the life I now constantly lead.' For a salary of £200, Watt worked on the canal three or four days a week, assisted by one overseer and by the contractors, whom he described as 'mere tyros'. There were more than 150 labourers at work, and there were frequent labour-disputes, particularly about the movement of workers from one job to another. Robert Mackell, the engineer of the Forth & Clyde, accused the Monkland contractors of taking his experienced navigators; and the contractors often took workers from one another, so that Watt had to introduce a regulation that any contractor who 'employed a man belonging to another' should forfeit 1s 'for each day that he wrought with him after he was claimed', and that any workman 'leaving his master and working with another without leave' should forfeit his wages. In December 1770 Watt wrote again to Dr Small: 'I have a hundred men at work just now, finishing a great hill we have wrought at this twelvemonth. The nastiness of our claygrounds is at present inconceivable—the quantities of rain have been beyond measure.' Watt, like Mackell, had numerous responsibilities: he had to survey, level, plan and stake out the canal, make bargains with the contractors and supervise their work, calculate the number of cubic yards excavated, check accounts and pay out wages. Some indication of his attitude to the role he had to assume is given by such bargains as that of November 1772, by which a contractor agreed to repair a tunnel under the canal 'for 7s 6d and a bottle of whisky'.[4]

The financial provision for the work was unsatisfactory, proprietors being frequently in arrears with their payments; and on one occasion Watt had to raise money on his personal bill in order to pay wages to his workers. Land prices, too, proved higher than he had expected. The road between Glasgow and Airdrie was further north than it is now, and Watt had to make a road from Airdrie to the head of the canal for the transportation of coal. The first water-supplies came from Airdrie South Burn, which was on almost the same level as the canal. On 28 July 1770 John Smeaton inspected the work and pointed out that the depth could be increased to 4 ft by banking, without any additional cutting; this suggestion was accepted three days later by a meeting of the proprietors, and was put into effect.[5]

1,000 yd west of Sheepford, a bridge had to be built to carry the Hamilton road over the canal; and west of the bridge was 'Muttonhole Cut', a cut 560 yd long 'through a great hill'. The bottom of the cut consisted of wet sand, and it was necessary to build wharf walls and then turf the slopes above; this work took almost a year. At the west end of Muttonhole Cut was the bridge carrying the Glasgow–Airdrie road over the Gartsherrie Burn. The canal was here taken over the burn by an aqueduct 16 ft wide and 72 ft long, and a new bridge was made to take the road over the canal. The cut through Drumpellier Moss presented difficulties, as the bottom of the canal sometimes rose 3 ft in a single night. The eastern part of the canal having been supplied with water from the Airdrie South Burn, the western part was supplied partly from Gartsherrie Burn and partly from a new reservoir constructed in Drumpellier Moss. Stone for the bridges came from quarries near the canal, and Watt designed boats to convey it; by 1772 he had six boats in use, the largest being 60 ft long, 10 ft wide and 4 ft deep. The completion dates of the bridges indicate the rate at which the work progressed: the Hamilton Road Bridge was finished in September 1770, the Aqueduct in April 1771, Coats Bridge in November 1771, the Gartsherrie Road Bridge in August 1771, Easterhouse Bridge in October 1771, Netherhouse Bridge in July 1772, Blairtummock (Bartiebeith) Bridge in September 1772, and Queenslie Bridge in October 1772. On 7 May 1773 Watt measured the whole canal and found it to be 7 miles and 15 chains in length; and soon after this work stopped on the canal, which had its terminus near Barlinnie Prison.[6]

In a letter to Dr Small Watt had explained that work was likely to stop 'from our having expended the subscription of £10,000 upon 7 miles of the navigation, and having about 2 miles yet to make'. By the Act of 1770 the proprietors had power to increase the capital to £15,000; but this was not done, probably because of the scarcity of money during the depression which followed the collapse of Douglas Heron & Company, the Ayr bank. In November 1772 Watt expressed the hope that the canal would prove profitable even in its unfinished state by virtue of its ability to undercut competitors; but this hope was not fulfilled, as maintenance costs at first exceeded income. It is said that when the proprietors were asked what should be done about the canal's finances, Mr

Colt of Gartsherrie replied, 'Conscience, lads, the best thing we can dae is for ilka ane o' us to fill up the sheuch on his ain lands, and let it staun.'[7]

In May 1780 it was agreed that a 10 per cent call should be made on the original subscription so that the affairs of the company could be 'extricated'; but this measure proved ineffective, and in June 1781 shares were advertised in the press at a price of £150 for a lot of five. Within a year all but eleven shares had been sold; and the remainder were then offered in two lots at £20 per share. The new proprietors extended the canal westwards to Blackhill and made a separate cut on a lower level to Castle Street; in July 1784 the *Scots Magazine* reported that a highwayman had been hanged in 'the castle yard', the traditional place of execution having been taken over for 'the termination of the Monkland Canal'. The principal shareholders were now Andrew Stirling of Drumpellier, the most important coal-owner on the canal's route, and William and George Stirling, partners in the Glasgow merchant firm of William Stirling & Sons. By 1790 these three men had become the sole proprietors; and by the end of the century they were said to have invested about £100,000 in improving the canal.[8]

The planning in 1785 of a set of locks at Blackhill to link the upper and lower reaches of the canal made it necessary to seek further supplies of water; and in the same year Whitworth and Millar raised the possibility of using the Monkland as an aqueduct to carry water into the Forth & Clyde. This could be achieved by extending the Monkland on the east to reach the River Calder near Woodhall and on the west to meet an extension of the Forth & Clyde's Glasgow branch. James Maxwell in his poem on the 'Great Canal' expatiated on the advantages which would result from this scheme:

> Behold, another scheme so well design'd,
> To have the great canal with Monkland join'd!
> This is a plan of such utility
> As no preceding age did ever see.
> Thus coals and other goods may here be brought,
> And carried to wherever they are sought.

The Act authorizing these developments was passed in 1790, and also gave the Monkland proprietors the right to widen, deepen and enlarge the canal where necessary, and to raise £10,000 above their existing stock.[9]

This Act and the threat of cheap coal from the Monkland collieries prompted the Glasgow coalmasters to unite in defence of their interests; but by 1793, when the improvements had been completed and Andrew Stirling at Faskine and Captain Christie at Dundyvan were mining 30,000 and 18,000 tons per annum respectively, the old monopoly had been broken. A description of Glasgow published in the following year described the canal 'to the village of Airdrie' as a work 'of the utmost consequence' which had brought to the city an 'enormous quantity' of coal 'equally requisite for domestic use and for the consumption of the fabrics'. It was hoped that the junction with the Forth & Clyde would permit Monkland coal to be exported through Bowling and Grangemouth; and in January 1794 the Governor and Council of the Forth & Clyde responded to a request from the Monkland proprietors by reducing their tolls on coal to ½d per ton per mile, except when water was scarce.[10]

The close relationship between the canal and the collieries, though generally beneficial to both, caused a serious accident in June 1791:

> About 4 o'clock in the morning, a very uncommon noise was heard immediately under the Monkland Canal in the land of Coats, when of a sudden the ground opened under the canal, and conducted the water into a coal-pit, where eight men were employed, two of whom escaped; but notwithstanding the utmost exertions of a number of workmen at hand, who were employed in damming or stopping the canal, they were unfortunately too late to save the remaining six men.

There was another serious accident in November 1795, when a storm raised the level of the canal so that it overflowed into the Gallowgate Burn, which consequently 'came down with such rapidity as to fill all the low houses in the Gallowgate, east side of the Saltmarket, and lower part of St Andrew's Square with water'.[11]

A more frequent subject of concern, however, was the lack of water; and there were frequent interchanges on this theme between the Monkland proprietors and those of the Forth & Clyde. In April 1794 it was agreed that the Monkland should be allowed to supply water to the Glasgow Royal Infirmary, as this was a 'useful and public establishment'; but in September the manager of the Monkland received an angry protest from the Forth & Clyde because he had drawn off water at an 'improper season of the year' in order to make repairs, thus interrupting the flow of water into the 'Great Canal' from the Black Loch. In the years 1796–8 work went ahead on a new reservoir at Hillend Moss, which was to supply the Forth & Clyde through the Monkland, and on the sluice and tunnel by which the Forth & Clyde were to convey water from the top to the bottom of the Blackhill locks; and in October 1798 the two companies made an agreement about the maintenance of navigation on the cut of junction during frost.[12]

In January 1804, after complaints about inequalities of depth in the cut of junction, it was agreed that the banks of the canal should be raised, provided the Forth & Clyde would raise those of the cut of junction and build a regulating-lock at the east end of Port Dundas. After the completion of this work in 1806 the minimum depth on both reaches was 5 ft. The total length of the canal was 12¼ miles, and there were two single locks at Sheepford with a rise of 21 ft and four staircase pairs at Blackhill with a rise of 96 ft. The width was 30 ft at surface and 15 ft at bottom. No dividend was paid until 1807, when the revenue was just under £5,000; but the improvements made by the Stirlings and the development of the coal and iron industry increased the canal's trade and brought the beginnings of prosperity. James Headrick, writing in 1813, reported that 60 to 80 tons of coal were conveyed to Glasgow every day by two-man boats which, when the wind was favourable, could let down a 'moveable Dutch keel' and hoist a square sail. Andrew, George and William Stirling were still sole owners of the canal; but the claims of Andrew's creditors made it necessary in 1813 to subdivide the stock and increase the number of shares from 100 to 2,020. The shares sold at this time were eagerly sought after; and by 1817 the dividend being paid was equivalent to £72 on each original £100 share, and the canal was sending 80,000 tons of coal per annum by the Forth & Clyde to Bowling. The revenue for 1822 was £14,544; and in 1830 the average annual revenue was said to be £12,000.[13]

The threat of railway competition appeared in 1824, when the Act authorizing the Monkland & Kirkintilloch was passed despite the canal company's opposition. The Forth & Clyde hoped to benefit from this development; but in 1828, when the railway had been in operation two years, they were still referring enviously to the Monkland's 'cent per cent' dividend. A proposal to carry the Garnkirk & Glasgow Railway under the Forth & Clyde near Hamiltonhill was successfully resisted by the Forth & Clyde because of the implied threat to the Port Dundas coal trade. Even as it was completed in 1831, running from near Gartsherrie to Port Dundas, this railway presented a direct challenge to the Monkland; but by reducing its dues the canal company was able to hold its own.[14]

It was helped in this by the establishment of new iron works in the Monkland district, and by the construction of side-cuts to the

Calder and Gartsherrie Iron Works, each of which was about a mile from the main line, and the Dundyvan and Langloan Iron Works, which lay about quarter of a mile from the main line in the parish of Old Monkland. The through coal trade to the Union Canal, however, was badly hit by competition from the Duke of Hamilton's collieries in West Lothian; and by 1834 it had virtually ceased. Passenger traffic was never of primary importance on the Monkland; but the passage-boats were described in 1827 as 'very commodious' and 'regular to their time' and were said to pass through a 'highly interesting portion of the country'. A boat started from the locks at Sheepford at 7.30 am and delivered its passengers at the Blackhill locks, from which another boat carried them to the Monkland Basin in Glasgow, arriving at 10 am; the service in the opposite direction left Glasgow at 4 pm and arrived at Sheepford at 6.30 pm, and the fares were 1s 'first cabin' and 8d 'second cabin'. After the *Swift* had been tried out on the canal on 9 July 1830, light boats pulled by two horses at 'upwards of 7 mph' were introduced; one of them was 85 ft long and 7 ft wide and carried 150–160 passengers. The total number of passengers carried rose from 11,470 in 1814 to 31,784 in 1834; it was remembered later that one of the boats had been called the *Defiance*.[15]

Agreement was reached with the Forth & Clyde in 1835 about the conversion of the Lily Loch into a reservoir; two-fifths of the cost was paid by the Forth & Clyde and three-fifths by the Monkland, and the work was finished in 1837. In 1841 a similar agreement was made for the raising of the existing reservoir at Hillend. The construction of the locks on the Monkland had been the subject of criticism since 1815; and in 1837 the company accepted James Leslie's suggestion that two new staircase pairs should be built alongside the two uppermost existing pairs at Blackhill, which were in a bad state of repair. Trade, however, increased so fast that it soon became evident that a single line of locks was inadequate; and in 1841 the two old upper pairs were rebuilt and two new pairs were constructed beside the old lower ones, so as to produce a complete double line. At the same time, the intermediate basins were enlarged, and a new graving-dock was built on the upper reach to replace one on the lower reach that had to be removed. Plans put forward by the Forth & Clyde in 1839 for improving Port Dundas included a further raising of the banks of the cut of junction; and after the Monkland had pointed out that this would involve their raising the banks, wharfs and bridges on their lower reach, it was agreed that the Forth & Clyde would take responsibility for any work of this kind that became necessary.

The proposed enlargement of the cut of junction was completed by 1843; and as a result Tennant's of St Rollox gave orders that their goods should be delivered by canal instead of by the Clyde. The canal company had incurred heavy debts through its work on the Blackhill locks; and in 1841 they obtained authority to raise £50,000 for the discharge of these debts and the maintenance of the canal, and to subdivide the existing 2,020 shares into 8,080.[16]

The Drumpellier Railway Act of 1843 authorized the construction of a short line from 'certain coalfields in the parishes of Old Monkland and Bothwell' to the Monkland near Cuilhill colliery. This line connected with a coalfield lying on both sides of the Calder and containing some 17,424,000 tons of coal. Its main purpose was to facilitate the transportation of coal and other minerals to Glasgow; and the proprietors included William and George Stirling. The line was to be 3,227 yd long, and was to consist of a 1,173-yd self-acting inclined plane from Rosehall to the Luggie Burn, a 1,576-yd inclined plane (worked by a stationary engine) from the Luggie Burn to the Edinburgh road, and a short locomotive railway from the top of this plane to the canal-bank. The cost was originally estimated at £26,000; but when £18,000 had been spent, a new estimate of £23,000 was given for the completion of the work. In the end, only the longer inclined plane and the locomotive line seem to have been made; but despite its incomplete state, the railway proved an important feeder for the canal, providing 900 boat-loads of coal per annum by 1849. In spite of railway competition, the canal company's revenue was said in 1845 to be £15,000 per annum; and the 'canal boat' was running twice a day, with the fares reduced to 6d cabin and 4d steerage. Money was still needed to pay off debts and to maintain and improve the canal; and in May 1846 the company obtained permission to increase tolls and to raise £40,000 by the creation of new shares. Negotiations had been taking place, however, on amalgamation with the Forth & Clyde; and after arguments about the relative value of the two companies' stock had been settled by an arbitrators' decision that three Monkland shares should be held equal to one Forth & Clyde share, the necessary Act was passed in July 1846. In August the officers and agents of the Monkland handed over their 'books and papers', and their steam tug was sold to the Ulster Canal Carrying Company for £660. The passenger-boats were sold to private operators, who continued to run them for a short time.[17]

A new depot was needed for the canal's pig-iron and coal trade; and in August 1846 Tennant's of St Rollox offered 4,000 acres at

18s per sq yd. A miners' strike reduced the canal's revenue from £10,520 for the half-year ending in September 1846 to £10,356 for the same period in 1847; but in the following year the trade was said to be 'steadily increasing'. The quantity of coal conveyed rose from 475,000 tons in the year ending September 1846 to 566,000 in that ending September 1848; and the canal also benefited from the demand for iron for railway construction. The growing trade placed a severe strain on the canal's water-supply, and in September and October 1849 it had to be closed for about six weeks. James Leslie and J. F. Bateman reported that no additional supplies could be obtained by enlarging existing reservoirs, and that the most practical measure would be either to pump water back from the lower to the upper reach at Blackhill or to adopt a plan suggested ten years earlier for constructing an inclined plane there. Since seven-eighths of the canal's trade was in coal and iron going downwards to Glasgow and only one-eighth in ironstone, limestone, manure and other goods going up to the Monklands, most of the boats ascending the locks were empty; and the company decided to construct a £6,500 inclined plane at Blackhill to handle these empty boats, hoping by this means to effect a saving of water which would enable the canal to operate satisfactorily for another decade on its existing supplies. This inclined plane, which was completed by August 1850 and in fact cost about £13,500, was 1,040 ft long and had a gradient of 1 in 10. The caissons in which the boats floated were 70 ft long, 13 ft 4 in wide and 2 ft 9 in deep, ran on two 7-ft railways, and were pulled by wire ropes 2 in in diameter, the motive power being supplied by two steam-engines of 25 hp each. Whereas it took 45 minutes to navigate the locks, the plane was able to pass boats in 6 minutes each. Between 20 March and 23 August 1851, 5,227 boats were taken up the incline and 225 sent down it; and it was used increasingly over the next 15 years, and not finally closed until about 1887. Glaswegians who had seen it in operation remembered it long afterwards as 'the gazoon'.[18]

Tonnage on the canal rose from 831,600 tons in 1846 to 1,058,310 tons in 1850, partly because the expansion of the shipbuilding industry on Clydeside was creating an increased demand for iron; but the construction of the inclined plane and the completion of the new reservoir at Roughrigg in 1852 relieved the water shortage. The Drumpellier Railway was bought by the Forth & Clyde in 1850, the price of £12,000 for the railway and a further £2,372 for land, sidings and the stationary engine being paid partly in cash and partly in canal company stock. In June

1850 the Bredisholm Coal Company proposed that the canal company should build a branch from the Drumpellier Railway to their pits; and it was agreed that when this had been done the canal company should receive from 8 to 10 per cent of the annual tolls, depending on the volume of traffic. This extension, which was 900 yd long, was completed in 1854, and it was hoped that further trade would result as 'new mineral fields in the direction of the railway were opened up'. In the same year consideration was given to the possibility of charging 1s per boat for the use of the Blackhill inclined plane; but the proposal seems to have been rejected. The collapse of an old tunnel over the Gartsherrie Burn near Coatbridge stopped traffic in April 1858; and in July the canal was closed again so that it could be cleaned (as was necessary every three or four years) and widened where 'the greatly increased up-traffic' had caused congestion. The 1790 stone bridge over the entrance to the old terminal basin in Glasgow was rebuilt at the same time, and the total cost of the work was over £900.[19]

Trade increased steadily from 1846 to 1866, as the coal and iron industries expanded in the Coatbridge area; and in the second half of this period the old mineral scows, which had carried 50–60 tons and had been operated by 1 boatman, 1 horseman and 1 horse apiece, were gradually replaced by screw steamers. Revenue from the Monkland for the half-year ending in April 1866 was down by £1,013, and there was also a drop of £373 in the receipts from the Drumpellier Railway. This was attributed partly to the 'unsettled working' of the Monkland miners, and partly to the opening of the Rutherglen and Coatbridge branch of the Caledonian Railway; but the most serious threat to the canal was the gradual exhaustion of the upper seams of coal along the route. It was recognized that prosperity could be maintained only if a new branch were made from Palacecraig near Sheepford to run 9 or 10 miles south into the expanding mineral field around Wishaw; but though this branch could have been constructed cheaply and without locks it would have presented a direct challenge to the North British Railway, and the canal company's proposal was consequently rejected by the House of Commons in October 1866. In the same month it was reported that the half-yearly revenue had fallen again by over £3,500 because of the miners' strike and the consequent stoppages in the iron works; and this decline continued in the following year. The canal was taken over with the Forth & Clyde by the Caledonian Railway in 1867, and thereafter little attempt was made to revive it. The Glasgow–

Coatbridge branch of the North British Railway, opened in 1870, took most of the canal's traffic between those places; and in the later nineteenth century many of the coal-pits on the canal were exhausted, as were the ironstone pits north-west of Glasgow which had hitherto sent ironstone to the Coatbridge works. The Drumpellier Railway was closed in 1896, its toll-receipts having fallen in 30 years from £2,953 to £721. By 1906, the canal's iron trade was dead, and its coal trade was limited to the supplying of a few works on the cut of junction, and of the electrical generating station at Port Dundas. The canal's revenue, which had been £35,214 in 1865, fell by 1913 to £1,915; and by 1925, despite a 50 per cent increase in rates, it had dropped to £351. Traffic stopped completely about 1935.[20]

In 1942 the London, Midland & Scottish Railway, which had inherited the canal from the Caledonian in 1923, applied to the Ministry of War Transport for permission to abandon it. Though it was no longer in use for transport purposes, the canal still supplied water to the power station and the Forth & Clyde, and permission was therefore refused. In 1943, however, the railway company reached agreement with Glasgow Corporation that if the closure were authorized the Blackhill locks should be filled in and the area around them transferred to corporation ownership; the railway company undertook to carry out the piping work necessary to maintain the flow of water to the Forth & Clyde. The Clyde Valley Regional Plan of 1946 described the canal as a 'formidable barrier' to communication, and said that while the 'amenities' were 'pleasant' in some sections, the canal in Coatbridge was 'alike an obstruction to development and a source of nuisance'. The disused locks at Blackhill, which lay immediately below Riddrie School, became notorious for drowning accidents; and in June 1948 1,000 parents petitioned Glasgow Corporation for action to end these. The canal was officially abandoned in 1950 by an order issued under the Railway and Canal Traffic Act of 1888; but the locks remained, and in June 1952 a Glasgow MP suggested to the British Transport Commission, who had acquired the canal four years earlier, that measures should be taken to prevent further drownings. In reply, the Commission pointed out that attempts to enclose the dangerous areas had run into difficulties, and instanced the removal of 750 sleepers from a fence on the Garncad–Townhead stretch.[21]

In December 1953 the cost of piping and filling-in the Monkland was estimated at £750,000; but it was agreed that some action would have to be taken, as the canal had become an 'eye-sore', a

V. Forth & Clyde Canal: (*above*) Camelon bridge about 1900; (*below*) the eastern entry, showing the river and the inner docks

VI. Forth & Clyde Canal: (*above*) a passage-boat at Port Dundas about 1840; (*below*) Glebe Street bridge on the Cut of Junction

'death-trap' and a 'breeding-ground for mosquitoes'. (The Monkland mosquitoes were known locally as 'flying tigers'.) Work began in July 1954 at the Castle Street Basin, and in the following month arrangements were made for piping the water at Blackhill locks. A deputation from Coatbridge Town Council to the 1955 Board of Survey pointed out that the canal was a serious obstacle to their development proposals, and that the cost of piping the water was beyond their means. The report, however, recommended that the canal should be maintained as a supplier of water; and its importance in this respect was enhanced in July 1960 when it was decided that Monkland water should be used at the new strip mill at Ravenscraig. In August 1961 Glasgow Corporation approved plans for converting some sections of the Monkland into 'modern highways', and authorized further piping and filling-in; and in July 1963 the government promised a 90 per cent grant towards the cost of piping and filling-in the canal in Coatbridge. Much of the Monkland has now been filled in, and its disappearance causes few regrets; but in its day it was the most profitable canal in Scotland, and made an important contribution to the industrial development of the Glasgow area.[22]

CHAPTER III

The Edinburgh & Glasgow Union Canal

DESPITE its grandiose title, the chief purpose of the Union Canal was not the establishment of a communication by inland waterway between Scotland's two major cities but rather the provision for the growing population of Edinburgh of an additional coal-supply comparable to that with which Glasgow had been provided by the Monkland. Some of Edinburgh's coal came from pits in the immediate vicinity of the city; but most of it came by water from Alloa, Wemyss and Newcastle, and was thus subject to a coastwise coal duty of 3s 6d per ton. The cost of land-carriage made it impracticable for the city to draw supplies from the Monkland coalfield; but it was hoped that a canal would solve the problem. In 1791 and 1792 discussions took place in Glasgow about a proposed canal from Leith 'by the south side of Shotts hill' into Lanarkshire, which was said to promise 'many important advantages'; and in Edinburgh in January 1793, there was a 'meeting of gentlemen interested in the communication by canal between Edinburgh and Glasgow and for the accommodation of Edinburgh and environs with coals at cheaper rate'. The Lord Provost of Edinburgh took the chair, and the 'spirit of the meeting' was such that 'the subscription was instantly filled': John Ainslie and Robert Whitworth Junior were commissioned to make a survey, and suggested four possible routes, all of them going from Leith to the Clyde at the Broomielaw. John Grieve and James Taylor then reported on the mineral resources on or near these routes, laying particular emphasis on the large coalfield in the Clyde valley. The Lord Provost hoped that the work, though 'of great magnitude and expense', would 'at no distant date be put into execution'; but the conflicting interests of landowners, coalmasters and manufacturers caused prolonged controversy over the route and thus prevented early action.[1]

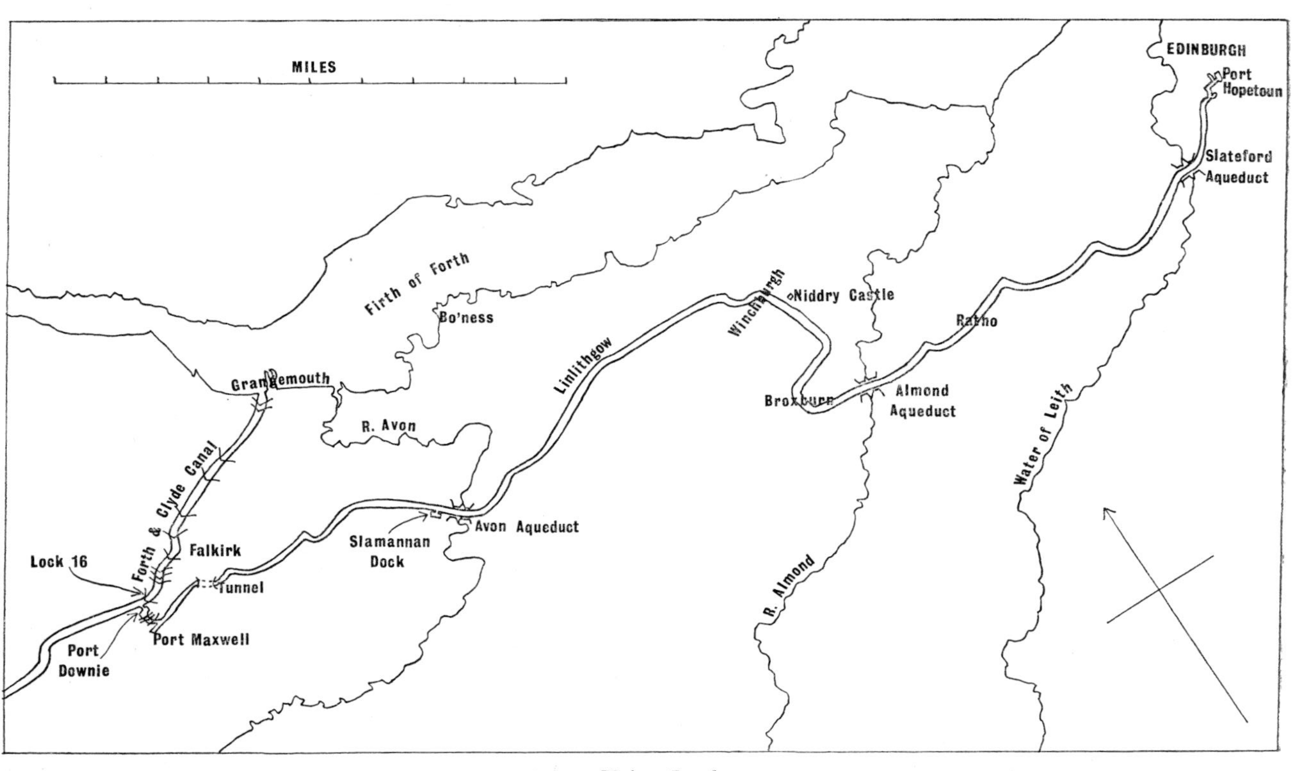

5. Union Canal

In 1797 John Rennie was asked to comment on the four lines surveyed by Ainslie and Whitworth, and proposed a fifth running further north; besides saving lockage, this route afforded the prospect of higher passage-boat revenue, since it passed close to Linlithgow and Falkirk. Rennie reported again in 1798, and from then on the lines seriously considered were his northern one, which ran by Ratho, Winchburgh, Linlithgow, Falkirk, Cumbernauld and Hillhead or Drumpellier, and what was called the 'Baton-moss line', which ran by Ratho, Midcalder, Baton-moss and Cleland. The Grieve-Taylor report was criticized by Henry Stewart, who declared coal-supplies to be equally plentiful further east; and John and Daniel Busby were asked to make another mineral survey. They concluded that there were 'inexhaustible' supplies of coal on the Baton-moss line; and Rennie therefore gave his support to that route, estimating the cost of a canal 27 ft 4 in wide at surface, 14 ft wide at bottom and 4 ft deep at £246,896. The war, however, was now diverting money and labour; and the project was shelved until 1813.[2]

In that year coal prices were unusually high in Edinburgh; and Hugh Baird, the engineer of the Forth & Clyde, was commissioned to draw up a new plan for the Edinburgh & Glasgow Union Canal. What was now suggested, however, was not a direct link from Leith to the Broomielaw but a branch from the Forth & Clyde at Falkirk to an Edinburgh basin in the Fountainbridge area, with a continuation past 'Mr Haig's distillery' into the Meadows. The chief functions of the canal would be to supply lime from Linlithgow and East Calder to the country between Ratho and Edinburgh, which was 'almost deprived of it', and to bring cheap coal to Edinburgh, which was paying one-third more for coal than Glasgow and three times as much as the area along the proposed route. There would be other cargoes, however, such as meal from the Avon and Almond mills and 'finest quality' freestone from the Brighton quarries. 'Neat, elegant and comfortable passage-boats' would be 'one of the greatest sources of revenue', since 'few travelled by regular stage-coach for comfort, ease or pleasure'. Baird estimated the cost of the canal at £235,167, and put the annual revenue at £52,727, including £20,000 from coal, £11,750 from lime, £3,000 from the passage-boats, £625 from freestone and £500 from timber and iron.[3]

The controversy over Baird's proposal was almost as violent as that over the 1767 scheme for a 'small canal' from Glasgow to Carronshore. Critics objected to the absence of any plan for communicating with the Broomielaw or with the harbour of Leith,

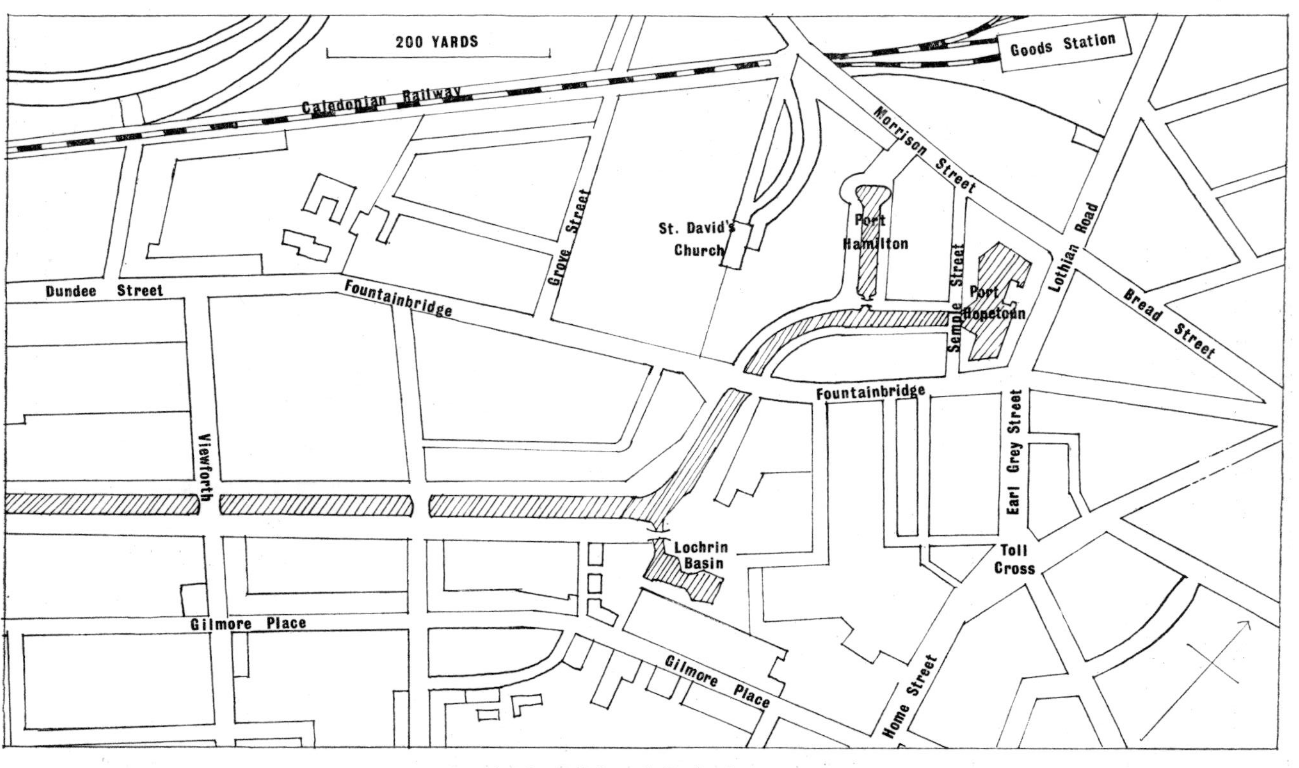

6. The Edinburgh Basins in 1901

and to the comparative lack of coal on the suggested route. It was said, too, that the scheme would bring some benefits to the Forth & Clyde on which Baird held 'a lucrative situation' and to the Banton Coal Works of which he was tenant, but be 'utterly useless to the public in general'. The route was also criticized because it passed through valuable land, beginning 'under the windows of one gentleman's house' and traversing 'shrubberies and pleasure grounds innumerable', and because it would be of no use for transporting coal to Glasgow, where prices were beginning to rise again as a result of a new 'understanding' among the coal-masters. Baird's estimates were also attacked, his allowance of £122 per acre for land being described as 'preposterous' and his hopes for high passage-boat revenue discounted, since the journey from Glasgow to Edinburgh would take twice as long as by coach, and few people would choose, 'for the sake of a few shillings, to do a voyage at once so uncomfortable and tedious'.[4]

James Grahame, in replying to these charges on behalf of the Union Committee, claimed that the plan was being supported by the magistrates and 'principal inhabitants' of Glasgow, Falkirk and Linlithgow, and by the Merchant Company and Incorporated Trades of Edinburgh, and opposed only by the monopolists of the Edinburgh coal trade, who were acting 'with all the violence and industry of terror and interest', knowing that there was an 'immense inexhaustible field of coal' south-east of Falkirk on Baird's route. The passage-boats between Port Dundas and Lock 16 were said to be 'cleaner than any inn in Scotland'; the passengers were 'sheltered from cold and rain in the cabins, and from the heat of the sun by awnings over the deck', and could 'indulge themselves with walking or sitting' instead of suffering the 'crowding and jolting, the cold in winter, the want of air under rain, and the heat and dust in summer' that attended a journey by stage-coach. The night-boats on the proposed canal would have 'commodious beds', so that passengers would be able to travel from one city to the other 'without fatigue or loss of time, and at the lowest possible expense'; the journey, by 'light improved boats', could be accomplished in as little as 9 hours.[5]

The subscribers for this scheme included General Maxwell of Parkhill and Alexander Livingstone of Parkhall, both of whom owned coal-pits near the line of the canal. Their opponents included the magistrates of Edinburgh, who in April 1814 declared their intention of opposing the Union Canal Bill, and appointed Robert Stevenson to report on the practicability of canals linking Lock 20 on the Forth & Clyde with Leith and Port Dundas with the Broomielaw. John Lauder and John Campbell, after a mineral survey of Baird's route, concluded that the coalfields west of the Avon where collieries were already open would, together with others at Callendar, Southfield and Bonny and between Bonnyhill and Castlecary, supply enough coal for '200 years more'. R. Bald made a mineral survey of Rennie's northern line, which had been revived as a rival to Baird's, and gave a favourable report: sandstone, limestone and ironstone were to be found west of Falkirk, ironstone, limestone and freestone between Cumbernauld and the Luggie, and coal in large quantities between the Luggie and the 'New Monkland Ridge'. John Paterson, in a detailed comparison of this line with Baird's, described the latter as a 'mere supplement' to the Forth & Clyde, designed primarily to serve the interests of Grangemouth. West of the point where they separated, the Union line would cost £23,140, whereas Rennie's would cost £63,690 if it went to Hillhead and £56,375 if it went to Drumpellier; but the additional expense would be fully justified by the extra revenue and by the 'vast facilities' afforded to commerce by a direct link between Edinburgh and Glasgow.[6]

In October Baird produced a supplementary report, raising his estimate for a canal 5 ft deep and 35 ft wide at surface to £246,321. In the following month the Union Committee decided to abandon the plan for a branch into the Meadows, since this gave the magistrates a pretext for their opposition to the bill; but they approved a scheme for linking the canal with South Queensferry by railway. In December Stevenson published his plan for a canal on one level from Princes Street to Port Dundas, with the possibility of locking down to Leith and the Broomielaw. The chief disadvantage of this plan was that it involved a 3-mile tunnel through Winchburgh Hill; but Stevenson conjectured that the canal-boats would be used primarily by those who 'consulted economy', and that half an hour in a tunnel would not therefore be an 'insuperable objection'. In the same month Rennie explained to the Committee his ambitious plan for taking branches from his 'level line' into the counties of Ayr, Lanark, Haddington, Berwick and Roxburgh, and persuaded them to accept his line from Bruntsfield to Lock 16 in preference to Baird's; but this decision was immediately reversed by a general meeting on the ground that the new line was 6 miles longer and altogether 'inexpedient'.[7]

Edinburgh had long suffered from lack of water; but the magistrates turned down an offer of supplies from the canal, saying that the canal water would be contaminated with 'filth poured in from crowded passage-boats'. The committee arranged for placards to be posted in Edinburgh, urging the population to petition Parliament in support of the Union Canal, which would provide 'cheap coal and fuel, an abundant supply of water gratuitously, and employment for thousands'. This had the effect of 'inflaming the public mind against the magistracy'; and when the windows of the Lord Provost's house were broken by a 'Jacobinical' mob enraged by his support for the Corn Laws, the supporters of the Union found themselves accused of stirring up revolutionary feeling in a city 'where the religious and moral principles of the population were perhaps better than any town in Europe'.[8]

Telford, who had been asked by the Union Committee to report on Baird's plan, declared in April 1815 that it 'would be the most perfect inland navigation between Edinburgh and Glasgow which the nature of the intermediate country could afford', all higher lines being 'inexpedient' and all lower ones 'impracticable'; he recommended, however, that the number of locks should be reduced by joining the Forth & Clyde not at Camelon but at Lock 20. Later in this month the second reading of the Union Canal Bill was moved by Kirkman Finlay, who attributed the opposition of a 'bare majority' of the Edinburgh Town Council to the lack of a continuation to Leith, but pointed out that such a continuation would be so costly as to make the undertaking impracticable. Opponents of the bill contended that it was 'calculated to serve the interests of the Forth & Clyde Company, not the public', and cited as an instance of the injustice being done to landowners on the route the fact that one gentleman was to be offered 'no other alternative than to approach his house by a drawbridge'. The Duke of Hamilton, who owned coal-pits on the proposed line, pointed out that when Rennie's scheme had been put forward 20 years earlier 'no person could be found to invest their money on a project hopeless of success'; but the bill was defeated by 33 votes.[9]

The supporters of Rennie's line met in July under the chairmanship of Sir John Marjoribanks, and opened a subscription for a canal from Leith via Edinburgh, Drumpellier and Glasgow to the Broomielaw; but only £52,000 was subscribed, and after a series of conferences between the two factions it was finally agreed in October 1816 that the Union line should after all be adopted. Baird prepared a new report, estimating the cost of the canal at £264,910 and the annual revenue at £49,000. Resistance was maintained, however, by the Wrights', Masons', Bakers' and Tailors' Incorporation of Leith, who argued that the proposed 'subsidiary cut' would perpetuate the Forth & Clyde's 'grating monopoly' and consolidate the advantages already enjoyed by

Grangemouth, and that passage-boats would be the only remunerative trade, since the cost of tolls and cartage would ensure that goods were still sent via the Forth. The magistrates having reversed their decision to refuse water-supplies from the canal, it became necessary for the Union Committee to refute suggestions that the water would be undrinkable by explaining their plan for a feeder from the Almond.[10]

The bill had its second reading in May 1817; and after Thomas Grahame and others had subscribed additional sums to satisfy a House of Lords standing order of 1813 which required four-fifths of the estimated expense of such an undertaking to be raised before the third reading, the Act was passed in June. It provided for a canal from 'near the city of Edinburgh' to join the Forth & Clyde 'at or near Lock Number 16, opposite to Camelon House', thus opening 'a direct, easy, expeditious and cheap conveyance for corn, coal, lime, manure, stone, timber, goods, wares and merchandise between the cities of Edinburgh and Glasgow, and to and from the adjacent towns and places'. The proprietors included the Lord Provost of Edinburgh, Alexander Livingstone, General Maxwell, and Robert Downie, who had made his fortune in India; and they were authorized to raise £240,500 initially and an additional £50,000 if necessary. The main line of the canal was to be 5 ft deep, but the section below the locks at Camelon was to be 10 ft deep like the Forth & Clyde. There were clauses restraining the company from using land belonging to William Forbes of Callendar, and from taking the canal close to such mansion-houses as Glenfuir. Authority was given for reservoirs on Cobbinshaw Bog and elsewhere, for feeders from the Almond and the Avon, and for water to be taken from a number of streams near the route. Tolls were to be 2d per ton per mile on limestone, building-stone, paving-stone, flagstone, coal, coke, culm, lime, bricks, tiles, slates, ore, dung, earth, sand, clay, peat, marl and manure, 3d per ton per mile on wood of all kinds, and 4d per ton per mile on corn and on 'all goods, wares, merchandise and things' not previously specified. Boats for conveying passengers and parcels could be licensed by the company for periods of up to 7 years. Edinburgh was to be supplied with 'waste water' when the depth exceeded 5 ft; and the Town Council was to be entitled to a duty of 1d per ton on all goods except manure which were loaded or unloaded within a mile of the eastern terminus.[11]

A general meeting in August elected a committee of eleven with Downie as chairman, and appointed Hugh Baird as engineer and George Moncrieff as clerk. Baird's salary was fixed at £500 per annum, provided the work was finished within 5 years for not more than £240,500; and he also received an initial payment of £1,500. The company adopted the motto 'Deruptione conjungo', made a 10 per cent call on capital, and arranged for the work to start after the harvest when more labour would be available. Offers were made to various landowners, including 3,000 guineas to Captain Maitland of Clifton Hall 'as a consideration for the inconvenience he may suffer in allowing the workpeople to come so near to his residence', and £450 per acre for land to be bought from the Trades Maiden Hospital. The work was divided into lots, and offers for these were invited. One of the contractors, Alexander McKenzie, was told to give a 'reasonable preference' to unemployed workmen who were 'natives of the place'. 'Strangers' from Ireland and elsewhere were employed, however; and two of these, William Burke and William Hare, later achieved notoriety in Edinburgh as the murderers who sold corpses to the Medical School. Complaints were received about workmen leaving unpaid debts; and Moncrieff was instructed to issue handbills 'warning all victuallers and others that they ought not to give credit without some guarantee'.[12]

Work was officially begun at the Edinburgh end in March 1818, Downie digging the first spadeful and throwing it into the air amid 'loud and continued cheering'. The ceremony was watched by 'a vast number of people', including many 'who for want of employment had left their houses, looking forward with the hope of providing for their families in a comfortable manner for some years to come'. Labour troubles arose during the summer, and the contractors were advised to maintain 'rigid economy' in their 'mode of living', to 'avoid company' and to persevere steadily 'in forwarding the work'. Williams and Hughes, to whom this advice was directed, reported in November 1818 that there had been 'riots' among their men, and that bagpipes had been heard in the middle of the night. Williams told the committee that the 'affray' had arisen from 'jealousy occasioned by a fight' some weeks earlier between a Highlander and an Irishman. The trouble had been aggravated by a public house at Niddrie, whose 'doors were open at all hours, Sunday not excepting, for giving drink to the men'. Twenty of the men involved had absconded, and assurances were given that none of these would be re-employed. The committee suggested that weekly instead of monthly payments might solve the problem, but Williams replied that this would produce weekly instead of monthly drinking-sessions; he agreed, however, to the setting-up of a savings-bank among his employees. In the following month, when some of his men were 'seized with fever', Williams suggested that it would be advantageous if they could be admitted to the Glasgow Infirmary without an application from the parish minister; and the committee therefore decided to follow the example of the Forth & Clyde by giving an annual subscription to the infirmary as long as the work lasted. Some time later, when the company were becoming afraid that the canal might lie 'unfinished and useless' as some others had done, Williams and Hughes were relieved of their contract because of their slow rate of progress.[13]

Discoveries made during the digging included 'an elephant's tusk entire', which for many years was preserved at Clifton Hall and 'shown to visitants with much politeness'. The country through which the canal ran contained many long-established estates; and many landowners, as Baird's critics had predicted, objected strongly to the presence of the canal workmen near their homes. The owner of Glenfuir House solved the problem by making the company buy his estate for 12,000 guineas; and since the house lay near Lock 16 discussions were held with the Forth & Clyde on the possibility of turning it into an inn to be run by the two companies in conjunction. These negotiations having broken down, the company resolved to convert the library into a bar and let the building to an innkeeper; but they had considerable difficulty in finding a tenant, and in the years 1823–9 the property brought in less than £100 per annum. An Act of May 1819 gave the company power to improve the route at several points; but they were still prohibited from crossing the property of William Forbes of Callendar. The need to avoid the Callendar estate forced them to take the canal through a tunnel just south of Falkirk; and when the agents for the estate insisted that the tunnel be built further west than had been intended it also became necessary to make a deep cutting for some distance beyond it. Difficulties arose in the building of the tunnel, which was to have a towpath and a waterway 13½ ft wide at surface; and the committee vented their annoyance in a bitter complaint about landowners who had charged excessive prices, demanded an unreasonable number of drains, bridges and roads, and forced the company to deviate at many points from the 'best and easiest' line.[14]

The tunnel, however, was not the only expensive work on the canal. The feeder from the Almond presented difficulties, as the banks of the river were extremely steep; and there were also the three great aqueducts over the Avon, the Almond, and the Water of Leith at Slateford. These were designed by Baird; but they were closely modelled on Telford's aqueduct at Chirk on the Ellesmere Canal, and Telford's advice was sought before they were built. In December 1818 it was reported that 'from numerous causes' the Avon and Slateford aqueducts were likely to be a 'very ruinous concern'; and in June 1819 Baird submitted new plans for the Almond Aqueduct, according to which it would be not a single 'large arch embankment' as originally suggested but a five-arch bridge constructed in the same way as the other two. Work on the Avon Aqueduct continued day and night during the summer of 1819; and in March 1820 it was reported that all three aqueducts were being built in an 'unusually substantial and improved manner' on the 'united advice' of Baird and Telford, and that the one at Slateford was 'superior perhaps to any aqueduct in the kingdom'. By August 1821, however, the improvements introduced to make the aqueducts more secure had increased the cost by £20,917. The heavy expense of these undertakings made it necessary for the company to obtain authority to raise another £50,000 on the security of the tolls; and the value of the shares fell greatly below par.[15]

While work continued on the canal, and on the private basin which was being made beside Haig's distillery at Lochrin, plans were being made for cargo- and passage-boats. The cost of the former was found to be considerable, and it was decided that only those who had experience from the Forth & Clyde should be employed as masters. The Forth & Clyde strongly recommended the adoption of iron boats, as these were more durable and easier to load than wooden ones; but the twelve boats which were being built in March 1822 seem none the less to have been of wood. For the passage-boats, the company looked to Holland for guidance: the clerk was instructed to obtain plans of the 'latest and most approved Dutch boats', and to have the regulations for the Dutch services translated at the company's expense. Baird was at work in November 1821 on the designing of a night-boat, which was to provide 'steerage accommodation', a 'good sitting cabin' and 'four sleeping cabins with four beds each': and by January 1822 the company's first passage-boat, an 'elegant and comfortable vessel' named the *Flora MacIvor*, was in operation between Edinburgh and Ratho. By March both the *Flora MacIvor* and its companion the *Di Vernon* had been let, and regulations for the service had been drawn up. The route was divided into four stages of almost 8 miles each, the intermediate points being Ratho, Winchburgh and Woodcockdale; and the time for the journey was estimated at 4½ hours. Breakfast and tea were to be provided, breakfast with two eggs costing 1s in the steerage and 1s 2d in the 'best cabin'. To prevent the introduction of anything 'offensive' into the cabins, the company prohibited smoking, the sale of spirits, and the cooking of any food other than eggs or potatoes, and gave orders that passengers in a 'state of intoxication' were to be put ashore, and that servants in livery were not to be admitted as cabin-passengers.[16]

Work on the canal continued at a feverish pace in the early months of 1822. There were press reports of the 'unusual exertions of the workmen', and of the 'line of blazing torches, by the help of which, even during the night, these exertions were continued'. The inhabitants of West Lothian and Stirlingshire, hearing of a survey of the line which the committee were to make in January by passage-boat, lined the banks and bridges; and 'the party in the boat, on entering Stirlingshire, were welcomed by an invitation to partake of the hospitalities of the county, and were, during the remaining distance, hailed by bands of music, flags, bonfires etc accompanied by the loudest cheers from the multitude of respectable people, who so crowded the surrounding high grounds and the canal banks, that the horses had not room to proceed'. The tunnel was not yet ready for water, and the passage-boat was therefore drawn through it on a temporary railway. At the far end it met the first two boat-loads of coal on their way to Edinburgh; and the committee decided to distribute 20 tons of this coal 'for the relief of the destitute sick'. The canal was officially opened in May, the first boat to make the complete journey by water bringing flagstones from Denny to Port Hopetoun. Stables for the track-horses and shelters for the passengers were begun in August; and by December the canal and most of the works connected with it were reported to be finished and in general use.[17]

The canal was 31½ miles long, 5 ft deep, 37 ft wide at surface and 20 ft wide at bottom. At the western end there were 11 locks, each 69 ft long and 12½ ft wide, which raised the canal 110 ft to a height of 242 ft above sea-level. The eastern terminus at Port Hopetoun had cellars, wharfs, stables, overseers' houses, shelters for passengers and an inn; and at the western end there was a basin known as Port Downie, which opened on to the Forth & Clyde just above Lock 16. The main water-supply came from the Cobbinshaw Reservoir, descending by the Bog Burn into the Almond and being taken near Midcalder into a feeder which met the canal just east of the Almond Aqueduct. The company's finances were far from sound: in March 1823 102 shares on which payments were in arrears were forfeited and put up for sale, and it proved necessary soon afterwards to obtain authority for the raising of further loans. £50,000 had been borrowed in exchequer bills in June 1820; and a further £50,000 was borrowed in the same form in June 1823.[18]

By the end of 1822 three committee members had built 21 barges at their own expense, and had let them 'on easy terms' to encourage trade; and in the following year, because of competition from the coaches, it was determined that dues should be reduced, and that a daily boat should be operated between Port Dundas and Port Hopetoun, so that canal carriage might be as 'certain and regular' as land carriage. It was assumed that the canal's main trade would be in coal; and the press anticipated a quick reduction in coal prices in Edinburgh after the opening. Dues on coal were fixed at first at 2d per ton per mile on short journeys and 1¼d per ton per mile on longer ones; but they were later reduced to 1d per ton per mile for all distances, with a 50 per cent discount by arrangement with the Forth & Clyde on coal shipped from Port Dundas to Port Hopetoun. The traders operating on the canal were subject to strict discipline, fines of £1 and £2 being imposed on those who passed the bridge at Linlithgow basin between 11 pm and 5 am, and severe punishment threatened for anyone who used coal-boats for carrying either passengers or smuggled whisky. Collieries from which coal was brought included the Duke of Hamilton's at Redding and Brighton, those at Bantaskine, Middlerig, Shieldhill and Banknock, and those on the Monkland; but at first they were only partly successful in competing with the Midlothian collieries, the Duke of Hamilton in particular being accused of fixing prices too high.[19]

Passage-boat fares between Edinburgh and Glasgow were fixed in 1822 by agreement with the Forth & Clyde. The cabin fare was 3s 9d for the Union section and 3s 3d for the Forth & Clyde section, and the corresponding figures for steerage passengers were 2s 8½d and 2s 3½d. These fares were reduced, however, in 1823. Facilities provided for passengers included an inn at Port Downie, 'cottages' at intermediate points, and a supply of books in the cabins. A guide-book published in 1823 gave a detailed account of the attractions of the journey, and especially of the three great aqueducts and the tunnel. The description of the Slateford Aqueduct, with its eight 'lofty' arches, was mainly technical; the author explained that the canal-bed, here and on the other aqueducts, was lined with iron and had a path and an iron balustrade on each side. In describing the Almond Aqueduct, on the other hand, he waxed lyrical over the 'wild charms of the sequestered spot' where it was situated; and he declared that the twelve-arched Avon Aqueduct was a 'noble edifice' which for 'magnificence' was 'scarcely equalled in Europe', and that the 'woody glens, the rugged heights and the beautiful Alpine scenery around' were certain to rouse 'sensations of pleasure in every feeling heart'. The great experience of the journey, however, was the passage through the tunnel, which was described in the language of a Gothic novel:

> When the passengers see the wide chasm, and the distant light glimmering through the lonely dark arch of nearly half a mile in length, they are struck with feelings of awe; and as they proceed through it, and see the damp roof above their heads, feel the chill rarified air, and hear every sound re-echoing through the gloomy cavern, their feelings are wound to the highest pitch.

Travellers were invited to regard this ordeal as the ideal preparation for the view across the valley from the top of the locks.[20]

In 1823 the canal was extended 570 yd westwards from the top of the locks, so as to reduce the minimum walking-distance between the summit-level and Port Downie; the terminus of the new cut was called Port Maxwell. There was comparatively little trade out of Edinburgh, except in 'merchant goods' and manure; but trade into Edinburgh increased rapidly, timber, stone, slate, brick, sand and lime being in great demand in the city, and coal being brought from '7 miles beyond Glasgow' and in considerable quantities from Redding and Brighton. To cater for this growing trade, the company laid out a coal basin just west of Semple Street, calling it Port Hamilton because of the Duke's 'great exertions' in supplying Edinburgh with coal. At Port Hopetoun itself a 'fine large building' was built for the 'luggage-boat companies' on the square where passengers landed, and other new buildings included offices, warehouses and dwelling-houses; boats bringing stone for these buildings from Redhall made an average of three trips a day. Boat-yards were established at Port Hopetoun and near Gilmore Place; and boat-builders in Leith were also kept busy by the canal. Some of these developments produced unfavourable comments from local residents; and in 1827 a Miss Dallas of Gilmore Place protested against the 'nuisance' caused by the keeper of the third drawbridge, who had erected a shed for cows 'with a boiler, a pigstye, and a shed for keeping their meat', thus spreading what the company's clerk described as 'somewhat of an offensive smell'.[21]

During the financial crisis of 1825–6 a number of traders told the company's clerk they were unable to pay their dues promptly, and the company decided that it would be unwise to adopt 'rigid measures'. The Exchequer Bill Commissioners refused to lend any more money; and instead £50,000 was borrowed from the Royal Bank, and an Act was obtained authorizing the allocation of the debt among the proprietors. Difficulties arose in 1832 over repayments to the Exchequer Bill Commissioners: the company had repaid £40,000 but were £20,000 in arrears with their annual repayments of £5,000, and the commissioners threatened to take possession of the canal. The company asked that the period of repayment should be extended from 20 to 24 years, but this was refused. In July 1833 it was stated that the interest on the two loans from the commissioners had been paid by the Commercial Bank.[22]

The problem of damage to the banks caused the company to begin experiments in 1829 with various lighter boats, including a small passage-boat which could be pulled by one horse. Experiments on the Forth & Clyde having shown that the wave following a steamboat was caused rather by its speed than by the operation of the paddles, and that a steamboat proceeding slowly made no more surge than one pulled by horses, John Neilson of Glasgow was asked to make a £754 steam-dredger for the canal; but when he failed to complete it on time the contract was declared void. Robert Ellis gave the committee an account of a two-horse gig-boat which carried 40 passengers at 8 mph on the Paisley Canal, and reported that the Forth & Clyde had ordered a light iron boat to carry 60 passengers; and it was agreed that Thomas Grahame should procure a boat of the latter type for the Union. This boat, the *Adelaide*, was brought from Glasgow in 1831; but in the following year it was found that the number of passengers had fallen, and that the fares were insufficient to cover running expenses. A new night-boat service for goods and passengers was planned in 1832. The boats were to be of light construction, and capable of making the journey from Glasgow to Edinburgh in 14 hours; and the fares were to be 5s cabin and 3s 6d steerage. Agreement was reached in September for the building of two such boats in conjunction with the Forth & Clyde; and in July 1833 it was said that they had 'lately begun to ply', and the managers were asked to ensure that the boys riding the horses were fit for their duty. At first, there were regular races between the night-boats of rival companies, but after complaints about injuries to passengers it was resolved that the times of departure should be altered, the boats of the London, Leith, Edinburgh & Glasgow Shipping Company leaving Port Hopetoun at 6 pm and 9 pm, and that of the Edinburgh & Glasgow Company leaving at 8 pm. The former company operated cattle-boats between Glasgow and Edinburgh,

VII. Monkland Canal: (*above*) Blackhill, with the locks on the left and the course of the inclined plane on the right; (*below*) the junctions of the Dundyvan and Gartsherrie branches about 1930

VIII. Union Canal: (*above*) Port Downie and the Union inn about 1900; (*below*) the north end of the tunnel about 1900

the fare for 'fat cattle' being not less than 1s 6d per head; and it was decided that revenue could be increased by admitting the 'lower classes' as passengers on these boats at a fare of 2s.[23]

On the main passenger services, efforts were made to reduce travelling-time, and notices were put up warning the public that the horses pulling the passage-boats travelled 'at great speed' and that it was therefore unsafe to walk on the towpath. In the spring and summer of 1835 six boats per day were leaving Port Hopetoun for Glasgow, the fares being 3s 6d cabin and 2s 6d steerage; and the company gave the boatmen and 'riding-boys' a dress allowance to ensure 'their more respectable appearance'. The number of passengers carried rose from 121,407 in 1834 to 127,292 in 1835; but competition from the Edinburgh–Falkirk coaches prompted the company to make a number of innovations. The timetable was adjusted so that passengers had 15 minutes for refreshments at Lock 16, and short-distance boats were established from Edinburgh to Ratho and Broxburn. It was agreed that travellers between Edinburgh and Linlithgow or Falkirk should be allowed to travel one way by day-boat and the other by night-boat at a fare of 3s cabin and 2s steerage each way for Linlithgow and 4s cabin and 3s steerage each way for Falkirk. Fares on the day-boats were fixed in December 1836 at 3s 6d cabin and 2s 6d steerage from Edinburgh to Falkirk and 7s cabin and 5s steerage from Edinburgh to Glasgow. Competition from the Edinburgh–Glasgow coaches caused some concern in 1838, a deficit of £756 on the passenger trade being reported for the first half of the year; and the committee decided to reduce their fares to 5s 6d cabin and 3s 6d steerage. Passage-boat revenue for this year was down by £1,491, including a £287 reduction in revenue from the night-boats; but in 1839 the revenue from the passenger trade was about £9,000, which left about £3,000 'clear profit'.[24]

Proposals to build a railway from Port Maxwell 'towards Glasgow' were 'taken into consideration' in 1830, but nothing came of them; and a proposal put forward by the Garnkirk & Glasgow Railway in 1835 for a line between Garnkirk and the head of the locks was successfully opposed, the company having discovered that the prospectus also suggested a railway to replace the canal between Edinburgh and Winchburgh. A rail link between Glasgow and the Union was established, however, in July 1840, with the opening of the 12-mile line of the Slamannan Railway from near Airdrie to Causewayend. This line, which had been authorized in 1835, terminated just west of the Avon Aqueduct at a basin 150 ft square which communicated immediately with the canal. It was hoped that the rail-canal route via Causewayend would now become the main passenger route between Glasgow and Edinburgh, and that the Union's passenger trade would benefit accordingly; and an omnibus-service between Port Hopetoun and Princes Street was maintained in expectation of increased trade. A special trip was organized to celebrate the opening, and the hope was expressed that as many committee members as possible would 'join the railway gentlemen at Glasgow in order to accompany them over the railway and canal'. In October, however, the committee's attention was drawn to an advertisement for a coach-service between Causewayend and Princes Street; and in December it was acknowledged that many passengers had been lost to the coaches. Difficulties developed with the railway company in the following month, first because the train failed to wait for the boat when it was a few minutes late, and then because the railway offices stopped selling tickets for the canal but continued to sell coach-tickets. The omnibus-service between Port Hopetoun and Princes Street was given up in January 1842; and in the following year, despite opposition from the Union, the Slamannan Railway was extended to link up with the Edinburgh & Glasgow.[25]

One bill for making a railway between Edinburgh and Glasgow was withdrawn in 1831; and John Lambert was sent to London to oppose another in 1832. A new campaign, attributed by the Union to 'feverish excitement among stock-jobbers chiefly in England', began in 1835; but after a struggle of 'almost unexampled length' the bill was rejected at the committee stage in 1837. In recognition of their efforts on this occasion, Robert Downie and Colonel Macdonald each received from the company a piece of plate worth 200 guineas; but the Edinburgh & Glasgow Railway Act was passed despite their opposition in 1838, and the line was opened in February 1842. The coach proprietors went out of business almost immediately, but the canal company tried to maintain its position by cutting fares and dues, and by improving its services. Dues on ale, cotton goods, cheese, cocoa, kelp, oil, raisins, soap, timber, saltpetre, whisky and pig-iron were reduced in April 1843; and in August experiments were made in the tracking of 8 loaded coal-boats by 4 horses. Two steam tugs with engines made by William Napier were ordered from John Wood & Company of Port Glasgow in December; and when the first of them was tried out the following April it was reported to be 'highly satisfactory'. In May 1844 fares between Edinburgh and Glasgow were reduced to 2s 2d cabin and 1s 4d steerage by day and 1s 4d cabin and 1s 2d steerage by night; and there were further reductions in July. In October, the company reported that the railway had originally established goods rates 25 per cent below those of the Forth & Clyde and Union Canals, and second- and third-class passenger fares equal to the canals' cabin and steerage fares; but they were confident that their reductions would frustrate the railway's attempt to secure a monopoly, provided the Forth & Clyde co-operated. New boats were still being ordered for the goods trade in 1845, and the tonnage of coal and stone carried was said to be increasing; but the battle for the passenger trade was already lost. The Union Inn at Port Downie lost most of its custom, so that the tenant had to sell her furniture to meet the arrears of rent; and in March 1848 the company decided to abandon its passage-boat services, though private firms continued for a short time to ply between Port Hopetoun and Broxburn.[26]

Negotiations took place in 1845 for the purchase of the Union Canal by the Forth & Clyde, which was at the same time to amalgamate with the Edinburgh & Glasgow Railway; the Union's debts were to be taken over by the Forth & Clyde, and the Union proprietors were to receive a fixed yearly dividend of 50s per share. The railway company, in explaining the scheme's advantages to its shareholders, declared that 'none could regret more than the directors themselves the continuation of the competition, or could be more anxious to put a stop to it'; and in anticipation of the Act arrangements were made between the railway company and the Forth & Clyde for the determination of their dividends by consultation, and between the Forth & Clyde and the Union for the payment of the proposed annuity and the reduction of the Union's dues. In November 1846, however, the Edinburgh & Glasgow withdrew from its agreement with the Forth & Clyde, thus ensuring that the amalgamation bill was lost in the House of Lords; and after considerable delay the Forth & Clyde, which had paid the annuity up to September, made it clear to the Union that no further payments would be made. In July there had been major breaches in the canal at Ratho and near the Avon Aqueduct, and the Union committee maintained that the Forth & Clyde, having assumed ownership of the canal, was responsible for the expense which had resulted from these. In reply to this, and to complaints about the money lost by the Union through the maintenance of low dues from September until the final decision about the annuity, the Forth & Clyde replied abruptly that they 'did not consider themselves under any legal obligation'.[27]

In October 1848 the Union established a working arrangement with the Edinburgh & Glasgow Railway; and in April 1849 agreement was reached on the terms of a bill for vesting the canal in the railway company. The Act, which was passed in June, pointed out that Edinburgh and Glasgow were linked by railway and 'enjoyed a communication by water by means of the Forth & Clyde Navigation and the Firth of Forth', and that the Union had paid no dividend since 1842, revenue having been sufficient only to maintain the works and pay interest on the company's debts. The company's capital stock was £240,500 in 4,810 shares of £50, and the cost of making the canal had exceeded this stock by £221,260. This debt had been distributed among the shareholders to the amount of £46 per share, and 3,174 shares had now been relieved of it, 1,636 remaining burdened. The company's total debt, inclusive of the amount thus allocated and not paid off, was £116,000; and its property was worth only £21,000. The railway was to raise the further £95,000 required to discharge the company's obligations; and in addition they were to create £114,000 of new railway stock, distributing £105,800 of this to proprietors of unburdened and £8,200 to proprietors of burdened canal shares, and pay £529 per annum to Edinburgh Corporation, which had been receiving about that amount annually from the canal company. Exclusive of this last undertaking, therefore, which was commuted to a cash payment in 1922, the Edinburgh & Glasgow paid £209,000 for the canal; and this was considered in later years to have been an excessive price.[28]

In 1865 the canal passed with the Edinburgh & Glasgow into the ownership of the North British Railway Company. Some attempt was made to promote trade, tolls being modified at intervals and special low rates being introduced for the carriage of coal from Causewayend to Port Hopetoun; but despite these measures there was a steady decline in the canal's income in the later nineteenth century. The total revenue was £5,209 in 1870, £5,247 in 1880, £3,775 in 1890 and £3,267 in 1900; and the evidence given to the Royal Commission in 1906–9 shows that the diminution of revenue was still continuing at the same rate, that the canal was 'weedy' and inadequately dredged, and that through traffic with the Forth & Clyde had virtually ceased. The final collapse of the canal's carrying trade came during its last years under the management of the North British:

	Number of Boats	*Total Revenue*	*Tonnage*
1907	56	£2,385	117,735
1921	32	£1,169	19,633

In 1912, when the Edinburgh slaughter-houses had been moved

out from Fountainbridge to Gorgie, the corporation raised the possibility of closing the now unnecessary basins of Port Hopetoun and Port Hamilton. Action on this proposal was delayed by the war, but in 1921 the basins were abandoned and sold to the city, the canal being shortened to finish on the south side of Fountainbridge. In 1923 the canal passed from the North British to the London & North Eastern; and in 1933, when commercial traffic finally ceased, work began on the filling-in of the locks and of Port Downie.[29]

After nationalization, there was growing pressure for the closing of the canal, which cost the British Transport Commission £4,000 per annum to maintain, hampered road transport by its low aqueduct over the A9 at Linlithgow, and presented a danger to children because of its proximity to housing in Edinburgh. It was now used only by a few pleasure boats; but it remained important as a supplier of water to industrial establishments in Falkirk, Linlithgow and Edinburgh, and in particular to the North British Rubber Works at Fountainbridge. In 1955 a petition with 3,000 signatures was presented to Edinburgh Corporation, asking them to press for the closure; and in a parliamentary debate shortly afterwards the members for West Lothian and Central Edinburgh drew attention to the problems created by the Linlithgow aqueduct and by the canal's attraction for children. The Board of Survey Report in the same year suggested that closure would make it possible to replace many of the canal's 72 bridges by culverts; and the 1958 Committee of Inquiry dismissed all possibility of reviving the canal as a waterway. Edinburgh Corporation gave its official backing to the campaign in April 1963, and the formal closure followed in August 1965. The closing of the canal has not, however, resulted in any significant change in its physical structure. From Falkirk railway station to the wharf at Fountainbridge it is still much as it was in its heyday—except that the Slamannan dock, where passage-boats and coaches once competed for the Glasgow–Edinburgh passenger trade, has become a nesting-place for swans.[30]

CHAPTER IV

The Glasgow, Paisley & Johnstone Canal

To the twelfth Earl of Eglinton, who owned large estates in the Ardrossan–Kilwinning–Saltcoats district, the eighteenth-century expansion of Glasgow and of its trade with America offered the opportunity to become the owner of an international port. Despite Golborne's success in the 1770s in deepening the channel, the navigation of the Clyde to the Broomielaw still presented difficulties for sailing-vessels; and the valley from Paisley to Kilbirnie provided the obvious route for a canal whose terminus on the Ayrshire coast would bear the same relationship to Glasgow that Liverpool bore to Manchester. In 1791, therefore, Eglinton formed a committee with the object of improving the harbour at Saltcoats and linking it with Glasgow by canal. In 1800, Rennie made a survey, plan and report on the harbour, and according to one account William Jessop was employed at the same time to report on the canal scheme. It was on Rennie's authority, however, that Eglinton recommended the canal to a meeting called by the Lord Provost of Glasgow in February 1803. £600 was then required for a survey, and since £250 had been raised in Ayrshire it was hoped that the rest would be contributed by Renfrewshire and Glasgow. The route was surveyed by John Ainslie, and he and Rennie reported in October 1804. The area through which the canal was to pass was said to be rich in minerals and to have 'many extensive manufactories'; and it was estimated that a canal for vessels of 25 tons would cost £130,960 and a canal for vessels of 60 tons £166,711. A meeting of 'noblemen and gentlemen' was immediately held in Glasgow, and £20,000 was subscribed; but it was resolved that, while Saltcoats would be linked to the canal by a branch line, the main canal should terminate in a new harbour to be established by Eglinton on Ardrossan Bay.[1]

A new report prepared by Telford and published in February

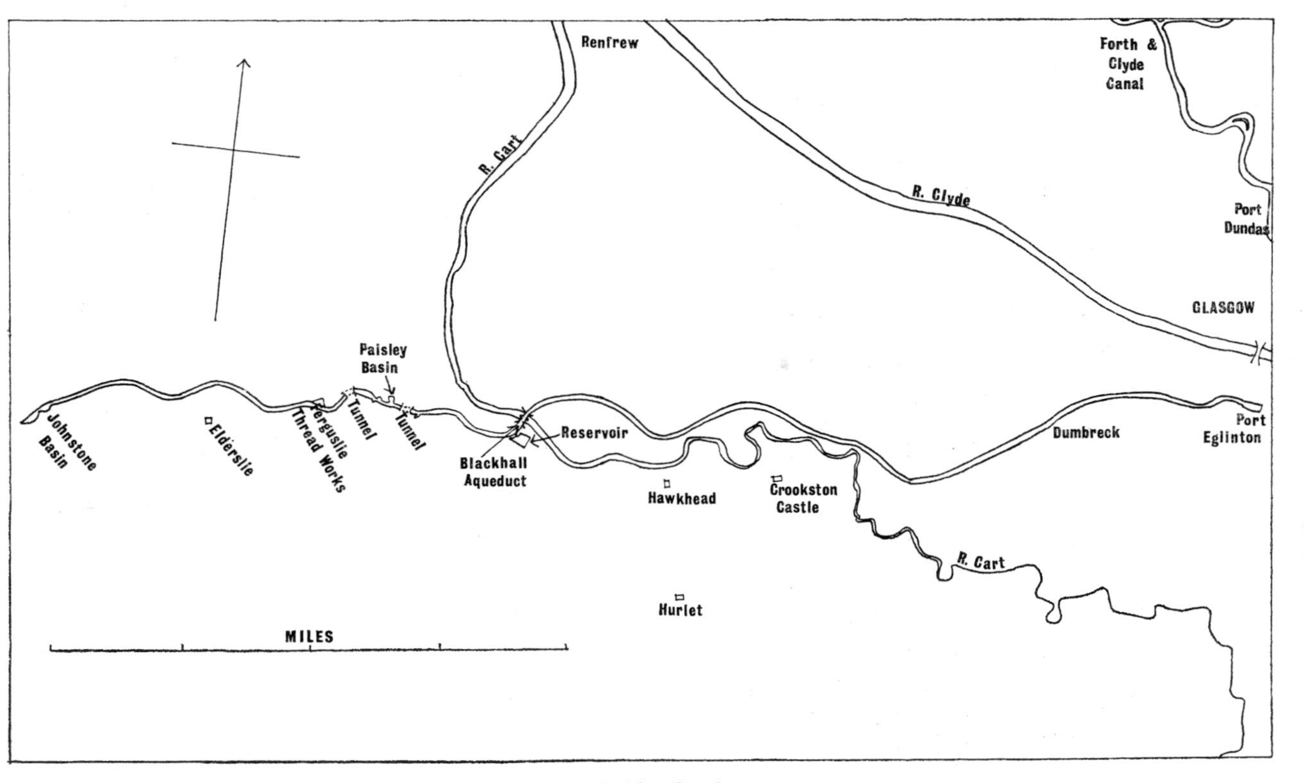

7. Paisley Canal

1805 not only incorporated this change but also modified the route between Glasgow and Johnstone so that the passage-boats in this densely-populated area would not be delayed by locks. A 'small lockage' west of Johnstone was to raise the canal to its summit level, which was to be maintained for 18 miles through a 'continual succession of coal, ironstone and limestone to within about 2 miles of the sea-coast'. The total distance by water from Tradeston to Ardrossan Bay would be about 32 miles; and it was estimated that the canal would cost £134,500 and bring in £13,699 per annum. Passage-boats were expected to yield a good return, and Telford believed the journey from Glasgow to Paisley could be done in 1½ hours, and that from Glasgow to Ardrossan Bay in 8 hours. The canal was to be connected with the Clyde Iron Works and the Hurlet Coal and Alum Works by branches 3¾ and 1¼ miles long respectively. A meeting held in Paisley in August 1805 adopted Telford's plan and decided to apply for an Act; and tribute was paid to Eglinton for promoting a 'measure of such importance to the general welfare of this part of the country'. The Act was passed in June 1806, and gave the company the right to make a canal from Tradeston to Ardrossan Bay with a branch to Hurlet and feeders to tap Ashgrove Loch and the rivers Garnock and Calf. £140,000 was to be raised in shares of £50, and a further £30,000 could be raised on the credit of the undertaking. The proprietors included the Earl of Eglinton and some of the partners in the Govan Coal Company; and the uses envisaged for the canal included the exporting of Scottish coal to Ireland, the importing of Irish grain for the populations of Glasgow and Paisley, and the cheap carriage of goods between those towns. Tolls were to be 2d per ton per mile on limestone, ironstone, building-stone, dung, earth, sand and clay, 3d per ton per mile on coal, coke, culm and lime, 4d per ton per mile on bricks, tiles, slates, ores and metals, and 5d per ton per mile on timber, bark, corn and grain.[2]

The subscribers met on 17 July 1806, and Eglinton was elected chairman. In November, it was resolved that the eastern section of the canal should be made first, since the 'great population and intercourse of that district of country' would ensure 'a considerable revenue' while the canal was being finished. Preparations continued during the winter, Telford being asked to draw up plans for the aqueduct over the Cart; and at the same time the 'ardent promoters' made great efforts to procure subscriptions. Charitable institutions were called on to contribute; and the City of Glasgow subscribed £1,000, the Trades House of Glasgow

£436 and the Tailors' Society of Glasgow £327. In May 1807, when £46,700 had been subscribed for 934 shares, work began between Johnstone and Tradeston. Some shares, however, were forfeited for non-payment of calls, so that the capital eventually stood at £44,342. Further sums of £13,348 and £57,860 were obtained by a loan raised among the proprietors and by an outside loan on the personal credit of members of the committee. There was little opposition to the canal from landowners; the most serious dispute was one with Alexander Spiers of Elderslie about the value of his land. The Earl of Glasgow pointed out that the canal would spoil the appearance of his grounds at Hawkhead, but was consoled by the thought that it would also increase the profits of his coal-mines. Robert Reid Cunningham feared that the water might endanger his mines at Stevenston; but his fears were of no practical significance, since the company's funds were entirely consumed by the construction of a canal from Glasgow to Johnstone.[3]

The section between Paisley and Johnstone was opened on 31 October 1810, and an 'elegant boat' began to ply on it. On 10 November, however, there was a serious accident at the basin in Paisley, when a fully-loaded boat newly arrived from Johnstone was overborne by the enthusiastic crowds leaving and boarding it, and threw about 100 people into the water. Press estimates of the number drowned ranged from 70 to 90, a large proportion of the victims being children who had been released from the cotton-mills for the quarterly fair. The boat righted itself immediately, and the passengers in the cabin were unharmed. 'A subscription immediately took place for the relief of the poorer class of those who had suffered by the loss of their relations.' The 11-mile canal from Glasgow to Johnstone was completed in 1811, and on 4 October the opening of navigation was celebrated by the Committee of Management and the Magistrates of Glasgow over 'a cold collation in the company's storehouse', and it was resolved that the Glasgow basin should be named Port Eglinton 'in honour of the patron of the undertaking'. From Port Eglinton, where an inn was built in 1812, the canal ran via Dumbreck to the north bank of the White Cart near Crookston Castle, where there was a small quay. The river was crossed at Blackhall by an 'elegant light aqueduct' 240 ft long and 30 ft high; and on its way through Paisley, where there was a small basin with a warehouse, the canal passed through a 240-ft tunnel under Causewayside and a 210-ft tunnel under Ralston Square. There was another basin at Johnstone, near where the Canal Garage now stands. The canal was 25 ft wide at surface, 13 ft wide at bottom, and 4 ft deep; and the bridges afforded 11 ft headroom. There were no locks; but the construction had involved a good deal of cutting and embankment. The depth was later increased to 4½ ft, and the surface width to 30 ft.[4]

Three passage-boats—the *Countess of Eglinton*, the *Countess of Glasgow* and the *Paisley*—were in operation by 1816. Each was 68 ft long and carried 120 passengers; and though the journey from Glasgow to Paisley could take up to 1½ hours, the slow speed had its advantages. 'Among the passengers', wrote James Cleland, 'the most advanced schools of religious and political opinion were represented, and the quiet, easy motion of the boat as it glided onward was favourable to the interchange of sentiment.' Attempts to raise funds for the original scheme were continued for some time. William Crossley having resurveyed the route and estimated the cost of the remaining work at £143,500, it was agreed in 1816 that Eglinton should apply for government aid. In support of the application it was stated that the canal would facilitate communication with Ireland, and that its construction would provide employment for a 'numerous body of labouring poor during the present stagnation of agricultural and manufacturing sources'. In May 1817, following the announcement that exchequer bills were to be used as a means of relieving unemployment, Eglinton applied to Lord Liverpool for a loan of £135,000; but his request was refused on the ground that the deepening of the Clyde under an Act of 1809 had rendered the canal unnecessary. Eglinton was still anxious to make Ardrossan a major port and secure cheap carriage for the coal from his mines at Kilwinning; and in July he declared that if the canal remained in its 'present unfinished state' his objects in subscribing for it would be defeated and he would consider himself entitled to be relieved of all his obligations. As late as 1820, the landowners of Renfrewshire were petitioning for government aid to complete the canal.[5]

The company's financial situation ruled out all prospect of completing the work without public aid. Loans were received in 1816 from Sir James Colquhoun and the 'Rev Dr Richardson', and Eglinton tried to persuade the Ayrshire subscribers to double their contributions. In June 1817 the committee tried to persuade shareholders to subscribe £50 per share in order to raise a sum which would relieve the obligants and reduce the canal debt to 'within a moderate compass'; but in September the debt stood at £70,911. By November the obligants had been forced to pay some creditors out of their own pockets: William Houston of Johnstone had paid £1,574, Robert Fulton of Hartfield £1,674 and Eglinton £1,721. The revenue for 1815, 1816 and 1817 was £3,044, £2,339 and £2,713 respectively; but a dispute arose as to whether this should be used to repay the company's debts or set aside for the continuation to Ardrossan. Eglinton and others brought a bill of suspension and interdict against the committee to prevent the company's revenue from being applied 'for any other purpose than the carrying on the canal'; and when this failed an attempt was made to fill the committee with Eglinton nominees. This scheme was frustrated by the other obligants at a general meeting of 6 November 1817, and it was resolved that the accumulated revenue should be kept *in medio* until the conflict between the obligants and the Ayrshire subscribers had been resolved. By March 1818 the obligants had advanced £27,000; but the debt continued to grow.[6]

In 1815, 1816 and 1817 the canal carried 22,570, 21,089 and 22,865 tons of goods respectively. New rates were introduced in December 1817, the charges for butter, oil, sugar, cotton and linen yarn, leather, iron, nails, slates, tiles and bricks passing between Glasgow and Johnstone being fixed at 5d or 5½d per cwt and that for grain at 8s 4d per ton. In 1819 it was reported that the 'common carriers' had reduced their rates and were securing much of the trade in cotton and other goods from the Johnstone mills. It was resolved, therefore, that the efficiency of the passage-boats should be improved so that they could complete the journey from Glasgow to Paisley in 1¾ hours, and that the Johnstone mill-owners should be informed that the company would undertake to carry their goods for 5 per cent less than they had been paying for carriage by road. In 1820 stables were built at Johnstone at a cost of £120 and the Earl of Glasgow opened a 2-mile railway from the canal to his colliery at Hurlet; and in the same year there were further changes in dues, the charge for grain between Glasgow and Johnstone being reduced to 7s 10d per ton. Competition from the carters continued; and in 1824 the company made a bid for the coal trade by offering 50 per cent discount for coal in excess of 10,000 tons sent by any colliery in a single year. An attempt was made later on to provide a twice-weekly service for through trade between Paisley and Edinburgh, the transference of goods between Port Eglinton and Port Dundas being effected by van. Great importance was attached to the traffic in cotton goods; and in 1836, after complaints from customers that samples sent from England had been delayed in transit and damaged by water, the company's Glasgow office was 'put under more efficient management' and a 'better system of communication' was established with the office in Paisley. In December 1837 the rates for cotton yarn and waste passing between Glasgow and Paisley were fixed at 5s and 5s 10d respectively.[7]

The number of passengers carried in 1817 was only 46,000; but in 1821 William Langmuir, one of the Paisley merchants on the committee, suggested that experiments should be made with swifter boats which would convince the public that canal travel was preferable to travel by coach. A plan for such a boat, estimated at £196, was studied in March; and in April 1822 it was determined that the company should attempt to still complaints about the tediousness of the journey by introducing a new boat 'of light construction' which could be towed at high speed. This boat came into operation later in the year, sailing daily from Paisley at 10.30 am and from Port Eglinton at 2.30 pm; and as a result, the passage-boat revenue for the subsequent twelve months showed an increase of £677. Fares on the passenger-service were fixed in 1830 at 1s between Glasgow and Johnstone, 9d between Glasgow and Paisley, and 3d between Paisley and Johnstone. In this year, on the recommendation of William Houston, the company made a series of experiments with 'gig-shaped' passage-boats about 30 ft in length. In April, a boat of this type with ten men on board was pulled 2 miles by a single horse in less than 10 minutes without raising any swell on the water. Two months later another boat was launched at Port Glasgow, towed up to the Broomielaw and transferred to Port Eglinton. Its trial took place on 4 June, a post-horse which had never towed a boat being substituted for the canal horse, which could not attain the required speed. The journey from Port Eglinton to Paisley was accomplished in 71 minutes; and with two post-horses and a lighter towing-line the return journey took only 45 minutes. The new boat was put in service immediately; and it proved so popular that the company decided in August to build an iron boat 70 ft long capable of plying both winter and summer and carrying 36–40 passengers.[8]

The company was recognized as a pioneer in the field of passenger traffic, and in 1835 the *Mechanics' Magazine* gave a detailed description of one of the boats used. This was 70 ft long, 6 ft wide and 1 ft 10 in deep, and drew 5½ in of water when empty and 19¼ in with a full load of 90 passengers. The weight of the ironwork was 17 cwt, and that of the finished boat 33 cwt. The cabin and steerage accommodation was covered with a 'cotton oiled cloth' supported by light curved ribs 2 ft apart fixed to the windows at the sides. Cabin accommodation was under the front part of this, and steerage accommodation under the rear part; and there was further seating accommodation fore and aft. In the bow and the stern there were decks beneath which light luggage could be stored; that in the bow was open to passengers, and that in the stern was used by the steersman. The rudder was 2 ft long and 20 in deep, and its bottom was in line with that of the keel. The cost of the iron-work was £70, and that of the whole boat £130. It was pulled by two horses, the first being blinkered and the second ridden by a boy. These horses were changed every 4 miles; and it was found that they could do 12 miles a day and still 'keep in excellent order', but that 16 miles was too much for them. Three sets of horses were employed for each boat; and since most of the bridges were of stone with towpaths through them it was not necessary, as on the Forth & Clyde, for the towing-lines to be detached.[9]

With the introduction of the new passenger-service, the company showed the utmost concern for the welfare of its horses. An additional stable was built at the Paisley basin, with a separate shed for sick horses; and after reports that the horses at Johnstone and Port Eglinton were 'very much neglected', orders were given that they should be looked after not by the boys but by the mates of the passage-boats. In September 1830 the captains were reminded that they were responsible for the 'proper driving and usage of horses at all times', and that the horses required attention on Sundays as well as weekdays. In May 1831 it was decided that stables should be built half-way between Glasgow and Paisley at East Henderston, as the horses were 'much distressed' by going the whole distance. In February 1834 additional accommodation was planned for the horses and grooms at Port Eglinton; and in April the groom at Paisley, who had more than thirty horses in his care, had his wages increased to 20s per week 'in consideration of his exclusive charge and long hours of attendance'. The wages of the passage-boat captains had been raised to 22s in March 1833 as an 'inducement to more faithful and conscientious discharge of duty', but this rate was maintained only while 'additional trips' were running. In July 1836 the boy riders struck for higher wages, and a boy engaged in place of one of the strikers was 'assaulted by one of the combination'. It was resolved that the strikers would be taken back only if their parents guaranteed their 'future good behaviour' and promised they would not leave the company's service without giving a month's notice. In October 1837 it was decided that the riders should be supplied with great-coats.[10]

The heyday of the passenger-service was in the 1830s. The

number of passengers carried rose from 79,455 in the year ending 30 September 1831 to 373,290 in the year ending 30 September 1835; and despite competition from the coaches and from a short-lived steam-carriage service on the Glasgow–Paisley road, traffic was maintained until the advent of railway competition in 1840 and 1841. A room at Port Eglinton Inn was let in November 1831 for the accommodation of the boat passengers; and in 1832 a shed was built for the gig-boat at Johnstone basin. In October 1833 there was an accident involving a boat carrying members of a masonic lodge to Glasgow for the laying of the foundation-stone of Telford's new bridge over the Clyde. The company disclaimed responsibility, since the accident had been caused by over-excitement among the passengers; but the safety-regulations were tightened in the following month, maximum dimensions being laid down for boats plying on the canal, and strict instructions being given that no boat was to be left in the track by day or by night. Fares were reduced in May 1835 'in consequence of the very great competition lately occasioned by the new coach company on the Glasgow road'. The cabin and steerage fares were now 10d and 7d between Glasgow and Johnstone, 6d and 4d between Glasgow and Paisley, and 4d and 2½d between Paisley and Johnstone. In December 1835 the company ordered an omnibus, which was to cost £110 and operate between Port Eglinton and the Trongate. In 1836 there were 8 boats daily in each direction between Glasgow and Johnstone; and in 1840 there were 7 daily between Paisley and Johnstone and 13 daily between Glasgow and Paisley. The demands of the passenger-service were such that the company at one stage employed a total of 78 horses. The shallowness of the canal, however, proved an obstacle to the introduction of steam-powered vessels; experiments with screw-propulsion were made in 1840, but were reported to have been unsuccessful.[11]

The poet and novelist Alexander Smith, who was brought up near Paisley, gives a nostalgic account of this period in *Alfred Hagart's Household*. He writes of walks along the canal, where 'no houses were to be seen', of the aqueduct which carried it across the river so that 'not a drop fell through', of the 'station house' half-way between Paisley and Glasgow where waiting passengers lounged about on the wooden wharf, and of the romantic view of Crookston Castle from the banks; and he describes the 'long white passage-boat with its horses and black-capped and scarlet-jacketed riders', identifies two of the horses as 'Smiler and Paddy from Cork', and tells how they were given a drink after being relieved and had the rest of a pailful of water thrown on their fetlocks. It seems clear that the passenger-service had earned a place in the affections of the local population.[12]

The plan for completing the canal to Ardrossan was dropped after 1820; but in 1825 and 1826 discussions were held about the possibility of continuing it by a railway. The financial crisis of these years made it difficult to obtain the necessary subscriptions; but in 1827 an Act was passed empowering the company to build a railway from Johnstone to Ardrossan and certain branch railways to link with it. Authority was given for £95,658 to be raised for the construction, this sum being arrived at by subtracting the £44,342 which had been subscribed for the canal from the £140,000 authorized in the original Act. Separate accounts were to be kept for the railway and the canal, and the railway was to be freed from responsibility for the canal's debts and was to distribute profits only to the new subscribers. In July 1827 the subscribers for the proposed railway announced that work was to begin on the section from Kilwinning to Ardrossan. Since the funds available were limited to £28,300 which had been subscribed and a further £20,000 which had been borrowed, the Ardrossan–Kilwinning section was the only part of the main line to be completed; but it satisfied the need (which the canal had failed to meet) for cheap transport from the Eglinton collieries, and proved highly successful for both the coal trade and the passenger trade. A sub-committee was appointed to manage the new railway, and in September 1828 James Sword was made superintendent with a salary of £120 per annum. In September 1829 Sir James Montgomery Cunningham offered to make a branch line from Saltcoats harbour to join the main railway near Saltcoats Toll-bar, and this offer was accepted; and in November 1833 he agreed to subscribe for 40 shares of the company's stock so that another branch could be made from his pits at Doura to join the main line near Dubbs. In November 1835 plans were made for lines to Fergushill and Irvine; but the second of these was never carried out, and by 1836 the company had accepted that the line from Kilwinning to Johnstone would also have to be given up. By an Act of 1840, therefore, the Glasgow, Paisley & Johnstone Canal Company and the Ardrossan Railway Company became wholly separate undertakings.[13]

Besides authorizing the railway, the 1827 Act resolved the conflict about the canal's accumulated revenue. After the expense of maintenance had been covered, this was to be used firstly in repaying the £57,860 plus interest owed to outside creditors, secondly in repaying with interest the £13,348 which had been raised in subscription loans, and thereafter in paying dividends on the original stock; and since there was no immediate prospect of funds being available for the second and third of these purposes, the effect of the Act was to place the management and revenue of the canal under the control of those proprietors who had been obligants for the outside loans, and to whom the first debt had consequently been transferred. In September 1831 and September 1834 these obligants received dividends of 2½ per cent and 7½ per cent on the sums they had advanced; and in September 1836 the sum of £5,756 was divided among them. Further dividends of 4 per cent, 7½ per cent, 3 per cent and 6½ per cent were paid to the obligants in 1837, 1838, 1839 and 1840; and by 1846 the preferable debt had been reduced to £62,000 including interest. Nothing, however, had been paid either to those who had contributed the the subscription loan or to the original subscribers.[14]

In 1831 there had been a proposal to build a railway from Glasgow to Johnstone to cater for the large passenger trade and speed up the carriage of cotton goods; but nothing had come of this. A line from Paisley to Renfrew was authorized in 1835 and opened two years later; but this presented a challenge not to the canal but to the River Cart. The first serious threat of railway competition came in 1836, when the bills for the Ayrshire and Greenock Railways were brought forward. A petition against them was prepared early in 1837, and in April William Houston went to London to organize opposition. Despite these efforts, however, both lines were authorized in 1837 and completed within a few years, the line from Glasgow via Paisley to Ayr being opened in 1840 and that from Glasgow via Paisley to Greenock in 1841. A period of 'ruinous competition' followed, both canal and railway companies carrying passengers between Glasgow and Paisley for 2d; and then in July 1843 the canal company agreed to give up carrying passengers and parcels in return for an annual payment of £1,367. The canal horses were put up for sale, and fifteen of them were bought by the Union Canal, whose passenger trade had still a few years to run.[15]

Because the line of the canal blocked the obvious routes into Glasgow from the south-west, there were numerous proposals for railways cutting across it. In 1839 the Glasgow, Paisley, Kilmarnock & Ayr Railway Company informed the canal company that they proposed to make a branch railway from Hillington to Barrhead, which would cross the canal 390 ft west of Rosehill bridge; and in 1845, when enthusiasm for railways was at its height, attempts were made by the same company and the Barrhead Direct

IX. Union Canal: (*above*) the Avon Aqueduct, looking east; (*below*) a bridge near Linlithgow

X. Union Canal: (*above*) the Almond Aqueduct; (*below*) the Slateford Aqueduct in 1823

Railway Company to persuade Parliament to agree to level crossings which would have made the canal unnavigable. In November of that year serious consideration was given to the possibility of selling the canal to the Glasgow, Paisley, Kilmarnock & Ayr Railway Company for £75,218; and it was suggested that if the sale did not receive parliamentary approval the railway company might be permitted for £37,500 to carry a branch over the canal at Port Eglinton and to make three more crossings between Paisley and the Blackhall Aqueduct. W. Barr, a lawyer and a canal shareholder, protested that this proposal was 'illegal and hurtful to the interests of the canal', and called on the shareholders to 'meet and resolve on joint measures to oppose it'. Despite the railway company's contention that the two companies when united would be able to give the public a cheaper and better service, the bill authorizing the sale was defeated, and the canal remained independent for another twenty years.[16]

The Glasgow, Paisley, Kilmarnock & Ayr Railway Company was taken over in 1850 by the Glasgow & South Western Railway Company; and in 1865 John McInnes, a Paisley lawyer and canal shareholder, wrote to George Lumsden, a Glasgow merchant, to recommend that the railway company should purchase the canal's preferable debt, postponed debt and original shares, giving in exchange as much of the railway company's newly-created preference stock as would yield the same income. In 1869 an Act was passed dissolving the canal company and vesting the canal in the Glasgow & South Western Railway. The railway company undertook to keep the canal open and navigable and to pay £3,471 annually out of the revenue, which was to be used firstly in paying the arrears of interest on the company's debts, secondly in clearing off the debts themselves and finally in paying dividends to the shareholders. The arrears of interest on the preferable debt amounted in 1868 to £1,265; and the arrears of interest on the postponed debt were £34,793. By 1873, the railway company had acquired all the preferable debt, £6,707 of the principal of the postponed debt, and 505 of the shares, and were willing to buy the rest of the shares at £4 each, which they said was 'rather more than the estimated present value', and the rest of the postponed debt and interest at 15s per £.[17]

Under the railway company's management the revenue continued to decline: in 1875 it was £2,542, and in 1880 £1,762. In 1881 the company drew up a list of the basins and wharfs between Glasgow and Johnstone which were still in use. These included the Paisley basin, where there was a mineral depot, the Blackhall wharf, which was used by the Gleniffer Soap Company, and the wharfs at Pollock Colliery and Saucel Distillery; but the majority were small wooden wharfs like those near Hawkhead Mains, and were used only for manure. In the same year, a bill was put forward for the closure of the canal: its supporters declared that it was 'almost impossible to describe a canal in a more thoroughly rotten condition financially and in a sanitary sense'. There was no serious opposition; and as soon as the Act had been passed work began on the laying of a relief railway along the canal's winding bed to connect with existing lines near Port Eglinton and between Paisley and Johnstone. The passage of time has destroyed all traces of the canal's basins in Glasgow and Johnstone; but the Blackhall Aqueduct survives as a railway bridge, and a considerable stretch of the canal has been preserved in the grounds of the Ferguslie Thread Works, Paisley.[18]

CHAPTER V

The Aberdeenshire Canal

IN 1793 certain 'noblemen and gentlemen' of Aberdeenshire united to finance a survey of the 'Don and Urie-side intended canal'. Their original plan was to make a canal up the Don to Monymusk, with a branch along the Urie to Insch; but when the subscription was opened it was for a canal from Aberdeen harbour to the junction of the two rivers at Inverurie. Advertisements began to appear in the *Aberdeen Journal* showing how the canal was expected to contribute to the prosperity of the county. One announcement stated that lands and 'valuable stone quarries' in the parish of Newhills were for sale: one of the quarries was 'the Dancing Cairn', which supplied the London market with paving-stones, and it was expected that the 'projected inland canal' would greatly increase its value. Another advertisement mentioned the proposed canal as an inducement to would-be buyers of the 'lands of Mugiemoss', through which it was to pass. A General Meeting of the subscribers was held in 1795, and it was agreed that work on the canal should begin as soon as the subscription reached £14,000. At this date the subscription was £11,000; and in the following year an appeal was made for those who were 'desirous of promoting the undertaking' to come forward with their subscriptions as soon as possible, so that their names might be inserted in the Act.[1]

The Act, which was passed in 1796, declared that the canal would 'promote the improvement and better cultivation of the inland parts of the country'. The proprietors, most of whom were landowners in the county, were authorized to raise £20,000 in £50 shares; no person was to hold less than one share or more than forty. If this sum proved insufficient £10,000 more might be raised by the admission of new subscribers or by mortgage of the tolls authorized to be collected; in fact, only £17,715 was subscribed. The *Aberdeen Journal* believed that the undertaking would be 'productive of the most beneficial consequences to the tract of country connected with it'. The cutting of the canal was carried out by several contractors, with John Rennie as consulting engineer and George Fletcher as resident engineer. The intending contractors were given an opportunity to see the line of the course on two days in March 1797; and in the same year meetings were held by the commissioners appointed by Parliament to settle differences between the owners of the ground and the proprietors of the canal company. A proposal by Rennie to enlarge the size of the canal caused some delay in the start of the work. He suggested that the canal should be 27⅓ ft wide at surface, 14 ft wide at bottom, and 4 ft deep; but when the canal was opened it was said to be '3½ ft deep, and 20 ft broad at surface water'.[2]

In the meantime, lands at Inverurie were being advertised for sale on the assumption that an increase in value was soon to take place, since the completion of the 'canal from Aberdeen' would result in the 'building of manufactories and other public works'. In October 1798 the company advertised for masons to build three locks and contractors to undertake the cutting and puddling of about 7 miles of the route. In the following year the final plan and survey of the Inverurie turnpike road was made, and advertisements for the sale of land in the town began to include the projected road as another attraction. Work on the canal was somewhat impeded by 'mischievous persons' who threw stones and rubbish into the cutting and otherwise damaged 'the works, the canal boats, and other implements thereto belonging'. A reward of £5 was offered for information leading to the conviction of the offenders; and a warning was issued that such offences were liable to a punishment of 7 years' transportation. In 1800 it became evident that more money would be needed. Calls of 10 per cent on each share had been made at intervals of a few months, and in April 1799 a call of 20 per cent was made. This reverted to 10 per cent in July of the same year. In the next year a General Meeting of proprietors was called 'for the special purpose of considering of the means of enlarging the funds of the company'; and two months later a 'special General Meeting' was called to consider a resolution in favour of borrowing money on the security of the tolls, the proprietors being asked to indicate 'to what extent, by way of loan, they were willing to go'.[3]

The company's efforts to solve the financial problem were unsuccessful and they appointed a Committee of Management, which was instructed to apply to Parliament for permission to raise an additional sum of £20,000 by the creation of new shares at £20. This second Act received the royal assent in March 1801. It stated that the company had cut a considerable part of the canal and had

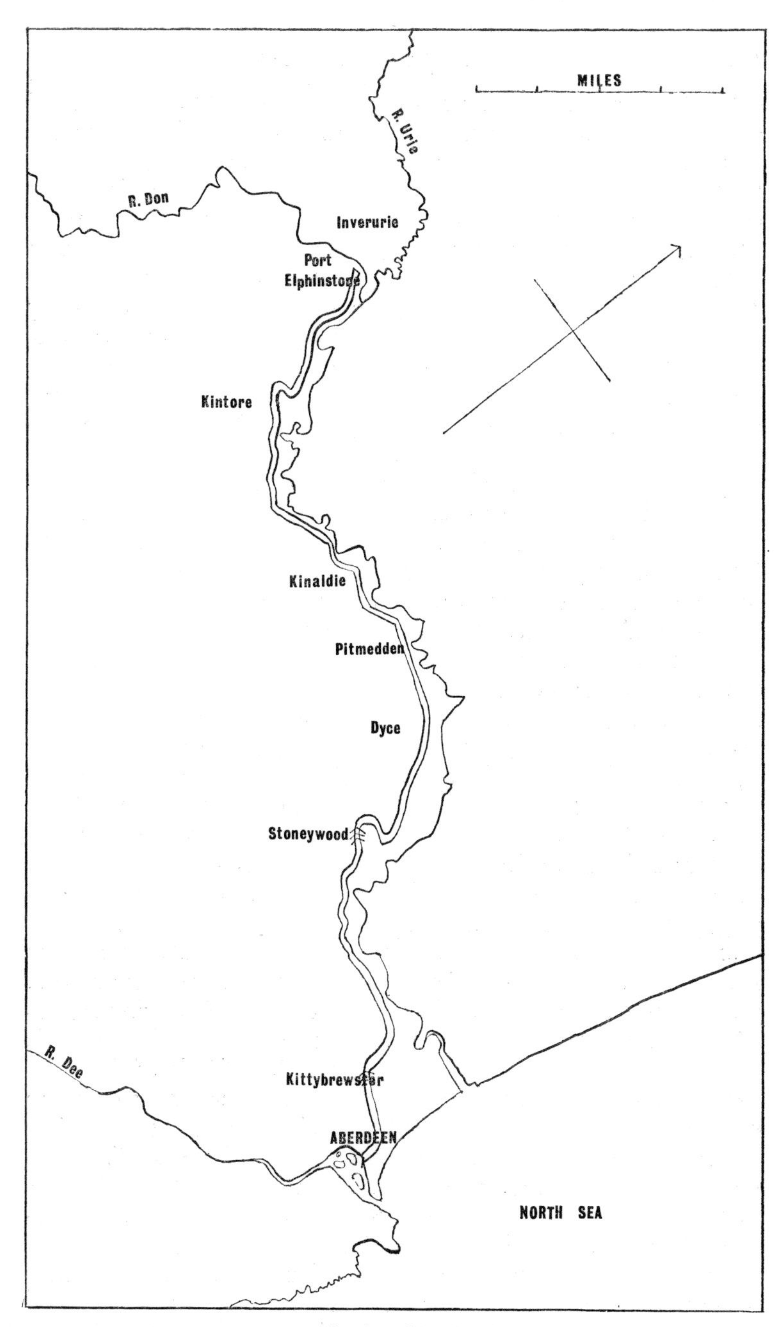

8. Aberdeenshire Canal

expended the whole of the money subscribed. They had incurred several debts, and attempts to raise further sums had failed. By the Act new shares of £20 each were to be issued. The owners of these shares were to receive a dividend of 20s per share, if the tolls of the canal produced this amount of profit, before the holders of the original shares received any dividend. When the dividend of 20s per share was reached, the original proprietors were to receive 50s per share; they were to be given an early opportunity to buy new shares up to the amount of capital they had already invested. The new shareholders were to have the same voting powers as the old. The original shareholders, however, did not take up all the new shares, and a 'public sale' was arranged in September; altogether the 1,000 shares fetched £11,421.[4]

As the work proceeded the hopes of landowners rose. In 1802 the sale of 'a lease and ground at Inverurie' was advertised as a very promising investment 'in view of the intended canal making Inverurie a sea-port to the upper part of the country'. Not everyone, however, welcomed the coming of the canal, and the contractors and workmen complained so frequently of being 'obstructed in digging and carrying away the necessary materials' that the Committee of Management, determined to stop these interruptions, reminded the public of some of the privileges and powers conferred on it by the Act of 1796. These included the right to supply the canal with water from any springs discovered in the process of its construction, from the River Don, and from all watercourses within 2,000 yd of the line, and to take stone for the works from any lands within the same distance. In February 1803 the company advertised for contractors to 'execute the cutting, puddling, masonry and carpenter-work of the canal from the Hangman's House to the harbour'. The Hangman's House was between the harbour and the city-centre, and it seems likely that this was the final section of the work; but in the following month a Special Meeting of the company was called to consider how this stretch of the canal might be completed, 'the funds of the company being insufficient for that effect'. The meeting was also to discuss 'what quantity of ground' would be needed 'for the basins, wharfs, etc both at the upper and lower ends of the canal', and to consider the possibility of mortgaging the 'tolls, rates and duties of the navigation' as a security for sums which might be subscribed or borrowed for the completion of the enterprise.[5]

Another problem was the settling of disputes over compensation for landowners. A meeting of the commissioners was called for 19 March so that a warrant could be issued to the Sheriff-

Depute of Aberdeenshire 'to summon a jury at the requisition of Alexander Innes of Pitmedden, on account of his alleged dissatisfaction with the sentence of the said commissioners'. In May 1803 tenders were again invited for 'that part of the undertaking lying between the Hangman's House and the harbour'. Various suggestions were made for raising extra funds; they included the creation of new shares, the issuing of annuities, and the leasing of the canal tolls. In February 1804 a General Meeting was held 'for the purpose of letting the tolls of the canal for 3 years from the time the same shall be navigable', and a Special General Meeting was announced for 1 May, 'for the purpose of borrowing the money requisite for completing the whole navigation, and granting mortgages, one or more, as security for the money to be borrowed', and for 'granting a lease of the tolls of the said canal to the highest offerer'. The sum of £10,000 was raised under a deed of mortgage of the tolls of the canal in favour of William Kennedy, who was granted a lease of the navigation for 3 years at £300 a year.[6]

At last in April 1805 the company declared that the canal 'from the basin at Quay of Aberdeen to the bridge over Don at Inverurie' would be ready 'at furthest by the first week of June'. Information about 'the low rates of tolls and duties for the tonnage and wharfage of all goods' was then provided 'for the encouragement of the public, particularly those who prefer the canal to land carriage'. The Act had laid down maximum rates or tolls for 'tonnage or wharfage of goods'. The rates were not to exceed 4d per ton per mile for 'hay, straw, dung, peat and peat ashes, and for all other ashes intended to be used for manure, and for all lime, chalk, marl, clay, sand, and all other articles intended to be used for manure and for all materials for the repair of roads', 5d per ton per mile for 'corn, flour, bark, wood-hops, coal, culm, coke, cinders, charcoal, iron, lime (except what shall be intended to be used for manure) stones, bricks, slates and tiles', or 6d per ton per mile for 'timber and other goods, wares or merchandise not hereinbefore specified'. 'Reasonable tolls' were to be paid for goods remaining on the wharfs above 48 hours. The tolls initially charged, however, were much lower than this, being only 1½d per ton per mile on stones and manure, 2d per ton per mile on corn, potatoes and coal, and 2¼d per ton per mile on most other goods, including timber. The announcement of these tolls was accompanied by that of a 'public roup' for the subletting of 'six new barges', which were to be used 'for the conveniencing of those who may not incline to furnish boats themselves.[7]

The opening of the canal took place early in June 1805, and

the *Aberdeen Jonrnal* gave a full and picturesque account of the celebrations:

> On Friday morning the Committee of Management assembled at the basin at Inverurie, attended by the Provost, Magistrates, minister and other inhabitants of that burgh, who congratulated them on the completion of an undertaking which must tend so much to the improvement of that and other parts of the country. The company then embarked on board one of the barges, *The Countess of Kintore*, handsomely decorated and fitted up by Captains Bruce and Freeman, and proceeded to Kintore, where they were met by the Magistrates and other inhabitants of that burgh. On their approach towards Aberdeen, they were joined by several parties of ladies, who were highly pleased with the novelty of the navigation through the locks; while several thousands of the inhabitants, crowding on the banks and bridges, added much to the interest of the scene. The company and a number of occasional visitors partook of refreshments on board the barge; and the voyage, which lasted seven hours and a half, terminated at the basin near the quay without the slightest interruption. The band of the Stirlingshire Militia met the barge several miles from town, and played many favourite tunes during the remainder of the voyage. The Committee afterwards dined together in the New Inn, when the health of the promoters of this great public work and every success to it was cordially drank. The canal passes about 19 miles into the interior of the country, rising 170 ft above the level of the basin at Aberdeen by means of 17 locks; and is 3½ ft deep and 20 ft broad at surface water. One barge has already delivered a cargo of coals at Inverurie, and another 85 bolls of shell lime at Kintore.

The construction of the canal involved no spectacular engineering feats. In addition to its 17 locks, it had 5 aqueducts, 56 bridges, and 20 culverts for carrying streams under the canal. The 17 locks, which were 57 ft long and 9 ft wide, were all within 4 miles of Aberdeen; and thus reduced the trade from the large granite quarries near the town.[8]

Three months after the opening the company advertised the auction of eight of its barges, 'separately, for the space of one month or longer period'. In the following year Sir Archibald Grant, one of the progressive landowners of Aberdeenshire, advertised a quantity of 'fir wood for sale at Monymusk', and pointed out that this wood was close to the Don, by which it could be floated 'to the head of the Aberdeen Canal'. Trade was seriously interrupted by the early collapse of the poorly-built locks. The *Aberdeen Journal* announced in October 1806 that the repairs were completed and the canal reopened. They 'sincerely' wished it success and had no doubt 'of its proving of great utility to the many spirited improvers in the district of the county through which it passed, by supplying them with coal and manure, so much wanted there, at an easy rate and with facility'. They added that the locks had been repaired 'in a very substantial manner' and that the canal had received the 'approbation of Mr Telford', who was 'eminently qualified to judge of the stability of such works'. The weather restricted traffic on the canal to 'the season', which was from 1 April to 1 December. In April 1807 the press announced the canal's reopening 'after a severe winter', adding that 'the spirited improvers in the Garioch' would 'soon experience the benefit to be derived from this inland navigation'.[9]

The first notice of passenger traffic was issued in July of the same year: 'A passage-boat, covered and neatly fitted up for the purpose of conveying passengers and light goods between Aberdeen and Inverurie, is now established upon the canal'. The boat ran three times a week from Aberdeen to Inverurie, and made the return journey the same day; the charge from Aberdeen to Kintore or Inverurie was 2s, and for any shorter distance the fare was 2d per mile. In the following September a daily service was provided, and a table of fares was advertised. The 'fore cabin', 'after cabin' and 'outside' fares from Aberdeen to Inverurie were now fixed at 2s 6d, 1s 6d and 1s, and there were corresponding fares for shorter journeys. In April 1808 it was announced that 'select parties' could be allotted cabins holding from 10 to 12 people if they gave a few days' notice to the shipmaster. The fare for a cabin 'up or down' was £2 2s, while that for the whole boat was £4 4s. The *Aberdeen Journal*, commenting on the progress of the canal wrote:

> Since the weather opened, the navigation has been unceasingly employed in conveying lime, coals, etc to all the inland parts of the county. Passage-boats have also been established, which ply daily to every distance on the canal, and afford a cheap and easy conveyance to farmers and country people coming to town on business. The agricultural interest in particular begin to be sensible of the utility of this Navigation, as facilitating in a most essential degree the improvements of the county, and affording to the more distant parts of it all the advantages of a contiguity to the coast and a ready market for the produce.

The canal's promising start did not, however, mean that the

shareholders were enjoying any profit. In September, in fact, a Special Meeting of the proprietors was held 'for the purpose of considering and authorizing a bill to be presented to Parliament to enable the company to borrow further sums of money for completing the works of the canal and paying off the debts presently affecting the undertaking'. The passenger traffic, however, was flourishing; and at the end of the season the company could claim, in advertising two fly-boats to let, that 'the great returns made to the former tacksman' had shown 'the decided preference given to this mode of conveyance'. The fly-boats (or gig-boats as they were also called) had preference over the barges at all times, and anyone breaking this regulation was liable to a penalty of 1s for the first offence, 2s 6d for the second, and 5s for the third.[10]

The final canal Act was obtained in 1809. It was needed because the company had incurred debts, and because 'sundry necessary works' still remained to be carried out. By this Act the company were empowered to raise a sum of £45,000 by promissory notes under the common seal of the proprietors. Interest was to be paid on them, and the holders were empowered to become shareholders 'in the ratio of their promissory notes'. The company was also given the power to raise money by mortgage of the rates authorized by the first Act, or to raise money by means of annuities. It seems probable that the company did not need to borrow these additional funds. No mention of this supplementary Act was made in the company's statement of 1848 on the winding up of its affairs, nor was there any reference to additional sums having been raised under its provisions.[11]

The value of the canal as an extra source of water was quickly discovered by the inhabitants of Aberdeenshire, much to the annoyance of the company, which was forced to issue a warning:

> Whereas of late the banks of the canal have been much broke and destroyed by many persons driving their cattle to water there, and others washing with tubs and other utensils on the edge of the water to the imminent danger of the canal breaking out and damaging the adjacent lands, Notice is hereby given that, if after this date any person or persons shall be found offending in these particulars, prosecutions will be immediately commenced against them for the penalties and damages in terms of law.

There were many cases of new-born children being drowned in the canal. One such case was reported in 1817: the body of a female child had been thrown in with a brick fastened to it 'with a view of sinking the poor innocent'.[12]

From October to December the passenger boats made only one journey a day. The company imposed a strict code of behaviour: no tipping was permitted, and no 'spirituous liquors' were allowed on board, and when it was incorrectly reported that one of the passengers had had his pocket picked and a pocket-book taken, the *Aberdeen Journal* contradicted this story, stating that the pocket-book had been inadvertently dropped into the water and that the person in charge, than whom no one could be 'more attentive or cautious in admitting improper company into his boat', had returned it. The fly-boats carried on successfully for a number of years, but the establishment of numerous coaches on the adjoining turnpike road, which offered speedier though more expensive travelling, began to diminish their profits. In 1816 the service of the 'Aberdeen and Inverurie Diligence' was advertised: it left Wood's Tavern in Aberdeen at 3 pm, and unlike the fly-boats continued to ply 'every lawful day during the winter months'. Robert Southey appreciated the view of the canal which he saw from the coach-window on his journey from Aberdeen to Oldmeldrum in August 1819. He described the canal as a 'losing concern to the subscribers' and a subject of complaint among the mill-owners on the Don, but a 'great benefit to the country, and no small ornament to it, with its clear water, its banks clothed with weedery, and its numerous locks and bridges, all picturesque objects and pleasing where you find little else to look at'.[13]

For the transport of bulky goods the canal was cheaper than the road; granite was one of the chief articles carried. In 1818 the *Aberdeen Journal* announced that 'a very extensive contract with the Government, for the supply of granite to the public works at Sheerness', had been awarded to 'some gentlemen in this place': the quantity required was about 700,000 cu ft, and this would 'give work to quarriers, labourers etc as well as afford employment to shipping for some time to come'. Probably because of this order 'immediate employment' was offered later the same year to carters 'to convey stone from the quarries near to Kintore to the canal wharf'. The Farmer Lime Company handled most of the canal's trade in lime and coal. In 1819 they stated that a 'regular supply of the best English lime and coals' would always be available at Port Elphinstone and Kintore. The grain trade on the canal received an impetus by the introduction of 'a new strong covered barge'. This, with ground on each side of the canal for granaries, was let in 1829. The owners of the barge claimed that 'this mode of conveyance' had long been wanted by merchants and grain dealers, and, 'there being now a grain market at Inverurie, granaries at the canal basin of Port Elphinstone would be of great advantage, as grain, as well as every other kind of goods', could be carried by the canal at 'a much cheaper rate' than by the road.[14]

Inverurie's position improved greatly as a result of the canal. In 1820 the Earl of Fife observed the 'thriving state of the burgh' and its increased population, which had trebled in a few years. Port Elphinstone or 'Inverurie Port' was described as a busy place which exhibited 'a scene not unlike the quays at Aberdeen'. At the beginning of the nineteenth century the population of Inverurie had been 'not above 500', but at the census of 1831 the number was 1,419, and in 1841 the figure had risen to 2,020, of whom 1,619 lived within the burgh. The advantages of Inverurie, as listed in a property advertisement of 1825, included the 'constant supply of coals at the canal head' and the canal fly-boats which ran 'twice each day during summer'. Kintore, 'the principal depot' for the canal, also increased in size: in 1811 the population was 863, and in 1841 it had risen to 1,299. The 'fore cabin' and 'after cabin' fares between Port Elphinstone and the 'fly-boat house' at Kittybrewster were cut in 1830 to 2s and 1s 6d; and the fares for shorter distances were 'reduced in proportion'. In 1840, the only year for which figures are available, there were 2,702 passengers northwards and 2,063 southwards by the less profitable of the two boats, and the fares collected totalled £299.[15]

In 1832 holders of the mortgage debt on the canal agreed to forgo their dividends for a period in order that the canal could be connected by a sea-lock with Aberdeen harbour. In the following year the *Aberdeen Journal* said that the sea-lock would be at the south end of the basin and that 'a draw-bridge' would be thrown across at the Lime Quay. This was sure to prove 'a very important and useful improvement' and one which would increase the revenue of the company. The sea-lock was built in 1834, and cost about £1,500. Its advantages were set out by the *Aberdeen Journal* in order to encourage its regular use:

> Grain dealers will get their grain conveyed at once from the country into the vessel, and thus be saved the expense of storing in town, and carriage to and from their granaries. Merchants may have their goods put on board a barge from the ship's side, and placed upon the wharf at Port Elphinstone, at little more expense than that of delivery in town; and country merchants sending butter, eggs, and other produce of the country, to be shipped from Aberdeen, may, in like manner, effect a large saving. Country distillers, by making Port Elphinstone their

> place of shipment, and taking a return draught of coals from thence, will not only save the expense of bonding in Aberdeen, but effect a very great saving in their charges for carriage. Stone merchants will save the whole expense of stone-stances at Aberdeen, besides being able to put the stones in the ship at a cheaper rate than they can now place them upon the stance; and a capital trade in slates may be established, by sending them as ballast to England in the lime and coal vessels. The rate of toll on all agricultural produce and on manure is very low, and a trade in potatoes and turnips may be very advantageously carried on.

An advertisement by the Aberdeen Lime Company, which used the canal as its chief means of transport, showed that they intended to export grain while continuing to import lime.[16]

The main commodities carried up the canal to Port Elphinstone in the years 1832–8 were lime, coal, dung, bark and bones, and the main commodities carried down to Aberdeen were oats, bear, meal, wheat, stones and slates; oats and bear, of which 36,430 quarters were carried in 1834, constituted approximately half the total tonnage. An iron boat was introduced about 1839; but the results were 'less favourable than might have been anticipated'. Abstracts of the accounts of the canal from 1840 to 1844 have been preserved. The main sources of income in those years were the Aberdeen and Port Elphinstone tolls, the fly-boat fares and the rents for land and houses. The major expenses were the cost of operating the fly-boats, the wages of lock-attenders and labourers, and the expense of superintendence and management. The gross profits for these 5 years were £470, £922, £1,335, £1,027 and £1,210. This money was paid out in dividends to the lessees and mortgagees. In 1844 the *Aberdeen Journal* announced that Aberdeen was at last to 'participate in all those advantages—social, commercial, and agricultural'—which 'invariably followed in the train of railway communication'. The proposed railway was to link Aberdeen with the south, but in the same year it was reported that the survey of the line northward to Inverness had made 'considerable progress'. This was to be known as the Great North of Scotland Railway, and in 1845 the secretaries of this company were instructed 'to obtain a statement of the present revenue of the Aberdeenshire Canal and such other information as might enable the directors to decide whether it is advisable to open a negotiation for the purchase of the canal'. The statement of receipts and payments for the 1844 season, which was duly sent to the railway company, showed an income of £1,659 from tolls and a profit of

£200 on the gig-boats; and an accompanying letter added that the revenue for 1845 would probably exceed that for the previous year by £200.[17]

The company decided to try to purchase the canal, but in case this proved impossible they worked out an alternative line which would go parallel to the canal 'within 50 to 100 yd of it all the way'. This plan was not needed, as the offer of £36,000 for the canal was accepted. The chairman of the railway company justified this 'large sum', which was 'about 26 years' purchase of the revenue', by saying that 'under all circumstances he considered the transaction desirable and advantageous to both parties'. The railway company undertook to take over the canal company's privileges and liabilities. This meant that they had to pay compensation to the tenants of the canal company, such as the Aberdeen Commercial Company, the Aberdeen Lime Company, and those individuals who had erected granaries, sheds, houses or wharfs with the intention of trading on the canal. There was, however, some delay in the payments to the canal company, and it was agreed that the railway company would pay the sum of £36,000 on 1 April 1848, on condition that the canal company was entitled to the whole revenue for that year and that the trade on the canal was not to be interrupted. The transfer of shares was to be made individually by each shareholder. In February 1848 the railway company, because of the 'unprecedented state of the money market', was unable to pay the agreed sum on the date arranged. The canal company accepted the situation, but reserved the right to call for payment at any time on giving one month's notice, and charged interest at the rate of 5½ per cent.[18]

In 1849 the *Aberdeen Journal* reported that a meeting of shareholders of the Great North of Scotland Railway Company had decided to lay single rails between Aberdeen and Keith, and that the estimated expense per mile, including the price to be paid for the canal, contrasted 'very favourably with the cost of the majority of lines'. Three years later it was stated that the Great North of Scotland Railway was about to construct the line between Aberdeen and Huntly, starting first with the line between Port Elphinstone and Huntly, so that the canal would be left open 'for the convenience of the public until absolutely required for the purposes of the railway'. In the same year an account was given of the cutting of the first turf for the railway at Westhall, 7 miles north-west of Inverurie. Sir James Elphinstone, the chairman, said that '60 years since, a namesake of his turned the first turf in commencing the works of the Aberdeenshire Canal', and expressed the hope that the railway would bring benefits 'as great as those which had resulted from the canal and the turnpike road'. The speeches at the ceremony included reminiscences of past transport, notably that of 'a poor man named Scorgie' who used to drive a 'caravan, with one horse' between Aberdeen and Huntly, but could not find enough passengers to make a living; and Sir James Elphinstone went on to give the chief reason for not starting from Aberdeen harbour and proceeding northwards: 'by carrying on the line to Huntly first, the public would still have the benefit of the Aberdeenshire Canal; thus the traffic already existing would be secured'.[19]

When the work on the railway had begun an unfortunate incident occurred. The transference of the canal to the railway company had not been completed when the contractors Erskine and Carstairs ordered the labourers to cut the canal bank near Kintore and set free the water. This dried up the section of the canal from Port Elphinstone to Stoneywood, and barges were suddenly grounded. At a meeting of the directors of the railway company a petition for interdict was presented 'at the instance of the canal company in consequence of an unauthorized interference by Messrs Erskine and Carstairs with the bed and banks of the canal'. Later in the same month the railway company declared that Erskine and Carstairs had been requested to stop operations, and promised that there would be no interference with the canal company's property. The canal was then repaired and refilled.[20]

Some of the canal proprietors expressed their willingness 'to take railway stock in part payment of the compensation to which they would be entitled in respect of their canal shares'; so the railway company, with the agreement of the canal company, decided to pay a deposit of £20,000, out of which sum bondholders' and shareholders' claims could be met. The railway company wanted public notice to be given that the canal would be shut 'about the first of January 1854', but in the middle of that month their contractors were still requesting 'definite instructions respecting the shutting up of the canal, and taking permanent possession of the same'. They received the reply that the matter was to be settled in 'ten days' and that notice was to be given to traders that it was to be shut. In February it was reported that £39,272—the agreed price plus accrued interest—had been deposited in the North of Scotland Bank for the purchase of the canal. In 1848 the balance of principal and interest due on the mortgage debt had been about £18,600, so the shareholders, who had never received any dividend, were unlikely to recover their capital; it was decided that no distinction was to be made between the holders of old shares and the holders of new shares. In March the canal traders sought valuers, 'mutually chosen without prejudice to the rights or pleas of either party', to value their boats; a shipbuilder and an iron-founder were accepted for the job.[21]

Finally, in September 1854, nine years after the project had been begun, the railway line was opened: 'An engine and two carriages, containing Sir J. D. H. Elphinstone, Sir Andrew Leith Hay, a number of other gentlemen, and several ladies, traversed the entire route of the line from Kittybrewster to Huntly. Everything along the whole line was found to be in a very satisfactory state.' The official opening came a few days later. The *Aberdeen Journal* gave a description of the course, mentioning Port Elphinstone as 'a place of considerable trade, caused by the terminus of the late canal being at the spot'. The arrival of the railway thus confounded Skene Keith, who had confidently predicted in 1811 that 'the competition between the canal and the turnpike roads on the banks of the Don forbade the attempting of an iron way in that direction'. Little now remains of the canal; but its track can still be traced at certain points, especially between Pitmedden and Kinaldie, where the railway took a less circuitous course.[22]

XI. Union Canal: (*above*) Edinburgh Castle from Port Hopetoun in 1825; (*below*) Port Hopetoun about 1900, showing the central warehouse

XII. Paisley Canal: (*above*) near Cardonald in 1866; (*below*) at Saucel about 1880

CHAPTER VI

The Crinan Canal

IN March 1771 the magistrates of Glasgow sent a petition to the Commissioners of Forfeited Estates asking that a survey should be made for a canal either from Loch Gilp on the west side of Loch Fyne to Loch Crinan on the Sound of Jura or from East Loch Tarbert on Loch Fyne to West Loch Tarbert. They said that if an entrance were made from the west sea into Loch Fyne by a canal of 7 ft or more in depth, half the herring busses could pass and re-pass that way without going round the dangerous Mull of Kintyre, and it would 'open easy and short communication between the Clyde and the west coast as far north as Cape Wrath and with all the Western Isles, so that timber, bark, kelp, grain and fish might be brought to market cheaper and with less danger than by doubling the Mull'. In June the same year, therefore, James Watt obtained leave of absence from the Monkland Canal to make a survey of 'Tarbit of Cantyre and Loch Gilp' for the Commissioners. Watt gave estimates for canals 7 ft and 10 ft deep for each route. For 7-ft and 10-ft canals from Loch Gilp to Loch Crinan the costs were estimated at £34,879 and £48,405 respectively; for canals across the isthmus of Tarbert the estimated costs were £17,988 and £23,884. The length of the Crinan Canal within the lochs was estimated at 4¾ miles, and that of the Tarbert Canal was put at 1 mile. The saving of miles by the Crinan passage was stated to be 75, and that by the Tarbert passage 55. George Clerk Maxwell, one of the Commissioners, examined Watt's survey and pronounced the work 'extremely accurate'. Watt did not directly express a preference for either route; but Maxwell considered that the Tarbert route ought to be 'entirely rejected' as it would, because of the narrowness of West Loch Tarbert, 'answer no good purpose'. No further action, however, was taken at this stage.[1]

Lord Napier recommended that public aid should be given to this project; and Dr Anderson, who was employed by the government in 1784 to examine the state of the fisheries on the west coast of Scotland, believed that a canal by the Crinan route was practicable, although he thought it should be constructed as 'a private undertaking for gain'. In 1785 John Knox described how he had sailed to Tarbert in order to examine the prospects for a canal there between Loch Fyne and the Atlantic. Tarbert did not 'fully answer' his expectations, so he hired a boat to Loch Gilp, 'where nature seemed to invite the public attention'. In his evidence before the Parliamentary Committee on British Fisheries in 1785, Knox said that a canal by the Crinan route would bring the people of the Highlands 'nearer to Glasgow and the seats of trade and commerce by more than 100 miles every voyage'. A year later the committee recommended Knox's proposal, accepting his view that the canal would 'greatly promote the fisheries and commerce of the north-western coasts of the island'. They concluded that as the work could be carried out at a moderate expense 'all reasonable encouragement' should be given to those who were willing to undertake it 'at their own cost'. This last recommendation must have disappointed Knox, who had hoped that the canal would be constructed by the government partly with public money.[2]

In 1789 Robert Frazer made a report for the British Fishery Society in which he outlined a plan for raising funds 'for opening a communication between Loch Fyne and the Atlantic through the valley that runs from Loch Gilp to Loch Crinan'. He believed that the Forth & Clyde Canal, which was nearly finished, would bring additional trade, and considered that Watt should be asked to make a new survey, because many 'improvements' had reduced the probable expense. The chief promoters of the canal were the Duke of Argyll, the Marquis of Lorne and the Earl of Breadalbane, the last of whom had interests in the slate-quarries on the island of Easdale. The Duke of Argyll and the Earl of Breadalbane employed John Rennie to make another survey, and in 1792 Rennie reported on two passages, one called the Daill passage and the other the Achnashelloch. Because of the shallowness of the water at the head of the lochs, Rennie had continued his survey along their shores; and the routes he proposed followed the coast-line for some distance on both sides of the isthmus. In each plan, the line of the canal terminated in the east at Ardrishaig on Loch Gilp; on the west the Daill line terminated at Portree (now known as Crinan) and the Achnashelloch at the point of Duntrune. The Daill passage was shorter by 2 miles, but the Achnashelloch was less expensive because it involved less cutting at the summit and because the cost of making the harbour was reduced. Both estimates were for a canal 12 ft deep, 30 ft wide at the bottom and 66

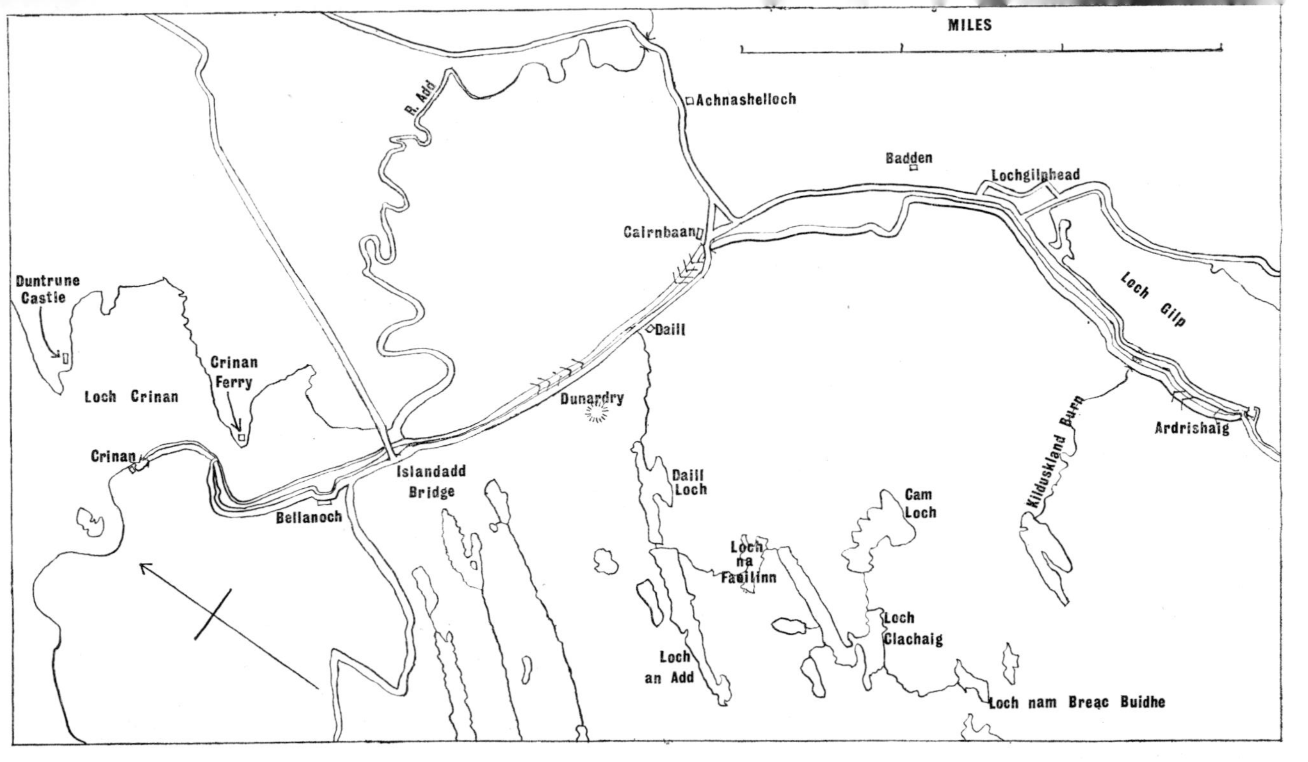

9. Crinan Canal

ft wide at surface. The estimate for the Daill passage was £63,628, that for the Achnashelloch passage £62,456.[3]

In their prospectus, the promoters explained the advantages of the canal, claiming that it would break down the isolation of north-west Scotland and the Hebrides by extending the fisheries and attracting vessels from the Clyde. The social benefits to the inhabitants of the west coast and the islands were stressed: it was claimed that the canal would bring coal from Glasgow, provide a market for fish and prevent emigration to America. It was hoped that this would persuade many to subscribe 'without an anxious regard to future profit'; but those who had some concern for a return on their money were assured that if one-third of the vessels in the coasting trade used the canal the profits were bound to be 'reasonable'. It was expected that there would be a great increase in the herring fishery and in the traffic in such goods as oatmeal, flour, grain, butter, cheese, woollen cloth, yarn, oil, hides and tallow and such local products as kelp, sand, slates, limestone, marble, quartz and timber. The promoters hoped, therefore, to see the subscription filled without application having to be made to the government for public aid.[4]

After Rennie's first estimates had been received, a meeting of subscribers was held in London in January 1793, and it was proposed that the canal should have a depth of 15 ft. The subscription list, which had been closed in December at £69,200, was reopened, since Rennie's estimate for a canal of the depth now suggested was £107,512. William Pulteney, alone among the subscribers, opposed the new plan; and 'finding a majority against him he endeavoured to frighten the English subscribers by saying that he considered the whole a losing concern and that it would not pay on any scale'. An 'English gentleman' replied that he and all those he had conversed with had subscribed 'entirely from the great public utility of the undertaking', wished to have it on the 'largest practicable scale to render it more extensively complete', and would be content if it answered that purpose, even if it paid a low interest. At this Pulteney withdrew from the meeting in 'high dudgeon'. Later the same month it was announced that £91,500 had been subscribed, and the proposal for a depth of 15 ft was then formally approved. The subscribers resolved in February 1793 that the Achnashelloch route should be followed; and it was agreed that a chairman and a small number of members should meet regularly at Inveraray Castle.[5]

The Duke of Argyll had already begun to make practical arrangements for the work, and in December 1792 had written to

James Ferrier to ask about the oats or meal which might be obtained in Liverpool. Supplies were scarce in Scotland, and he thought a supply of provisions might entice Highlanders 'from a great distance', whereas if they failed to obtain food they would from 'mere necessity engage themselves in some of the numerous works of the same kind going on in England'. The Duke of Argyll suggested the construction of 'huts for lodging the workmen', offering timber from the thinning of his fir plantation at a 'very low rate'. In the same month the committee asked Rennie to superintend the work, to provide iron tools, carts, wheel-barrows and 'other wheel carriages', coal, meal, stone, sand, clay, timber, iron, lead and huts at 'different stations', and to treat with the landowners for their ground.[6]

At another meeting held in February 1793 the decision to make the canal 15 ft deep was endorsed, and the subscription was said to have reached £107,700. In May an Act authorizing the canal from Loch Gilp to Loch Crinan was obtained. The proprietors included the Duke of Argyll, the Marquis of Tweeddale, the Marquis of Lorne, the Earl of Breadalbane, Lord Frederick Campbell, Sir Archibald Edmonstone, William Pulteney, David Dale, John Rennie and Josiah Wedgwood. They were authorized to raise £120,000, and a further £30,000 if necessary. The canal was to run from Loch Gilp at or near Ardrishaig to an unspecified point on Loch Crinan; materials could be obtained from the islands of Arran, Bute, Jura, Islay, Mull, Colonsay and Gigha, from the Cumbraes, and from parts of the Argyll mainland. Tolls were not to exceed 3d per ton per mile on any goods carried on the canal; and those on coal, salt outward bound, lime, limestone, shell sand, marl and manure were not to exceed 2d.[7]

English capital was an important factor in the original subscription: of the £50 shares on which payments were to be completed, 1,378 were taken up in England as against 473 in Scotland. The list of shareholders, however, included the City of Glasgow, the Glasgow Chamber of Commerce and more than forty Campbells. General Meetings were held in London; and in May 1793 the Duke of Argyll was elected governor of the company, Robert Campbell was elected chairman of the committee of management, and John Rennie was appointed chief engineer at a salary of 200 guineas per annum. Captain Joseph Huddart, one of the most famous marine surveyors of the day, was asked to make a survey of Loch Gilp and Loch Crinan and report his opinion on the best situations for harbours; and in December the committee of management was authorized to enter into contracts for cutting

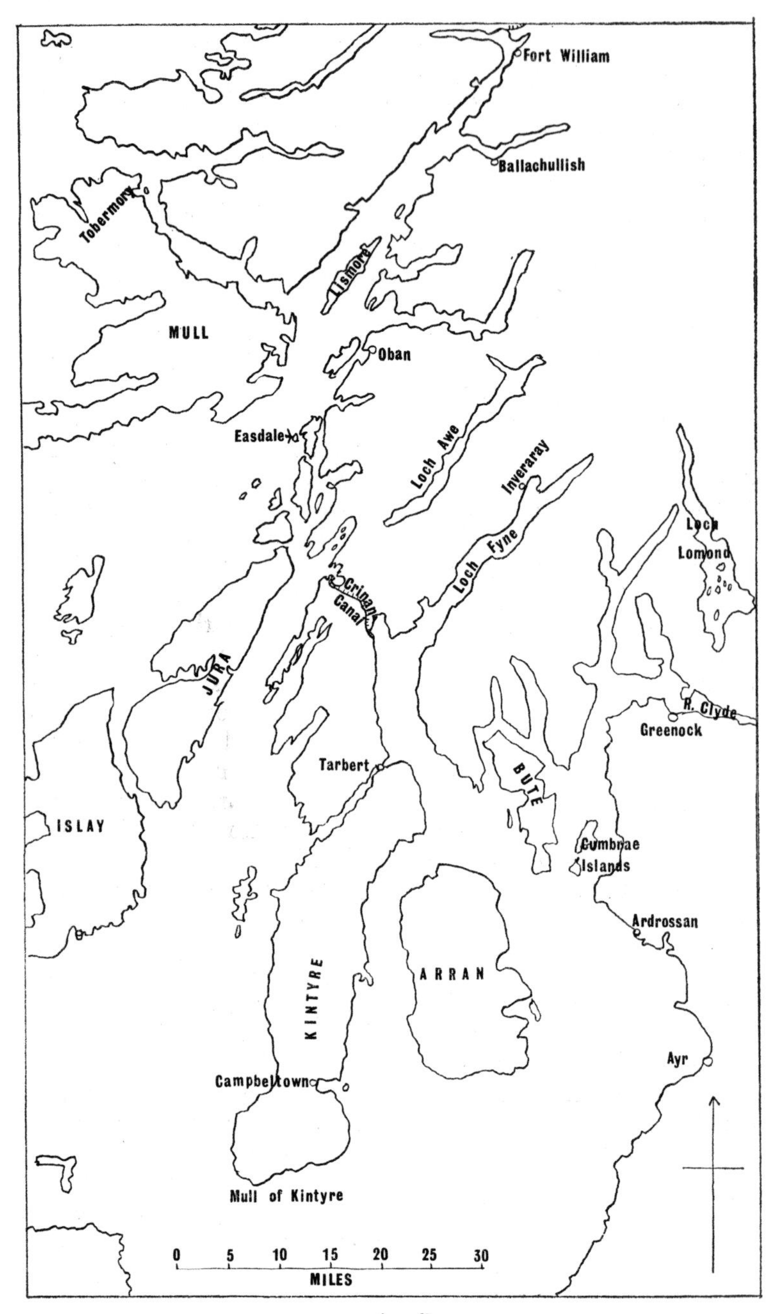
Fort William
Ballachullish
Tobermory
Lismore
MULL
Oban
Easdale
Loch Awe
Inveraray
Loch
Lomond
Loch Fyne
Crinan
Canal
JURA
R. Clyde
Greenock
Tarbert
BUTE
ISLAY
Cumbrae
Islands
Ardrossan
ARRAN
KINTYRE
Ayr
Campbeltown
Mull of Kintyre
0
5
10
15
20
25
30
MILES

10. Argyll

the canal and for the masonry. In February 1794 Rennie was empowered to procure a steam-engine for removing earth, and there was some discussion about the depth of the canal. The committee of management were in favour of a canal for vessels drawing 12 ft of water, as this would afford the proprietors a better prospect of profit; but Rennie and Captain Huddart were 'inclined to think' that the 'present plan of 15 ft' was preferable, since the 'principal part of trade from the east sea' would be excluded from passing the smaller canal. A little later, however, Rennie came down in favour of a canal 12 ft deep; and a decision to make the canal of this depth was taken in January 1795. It was determined at the same time that the Daill route should be followed after all, and that the locks should be 96 ft long and 24 ft wide.[8]

Financial difficulties arose very early in the canal's history, chiefly because of the failure of some of the proprietors to pay up their subscriptions. A few, like Joseph Inkersole of Leicester, asked for more time to pay; others wanted to withdraw their subscriptions, although these requests were refused. In May 1794 it was resolved that all who were in arrears should be charged interest; and in December it was agreed that those proprietors who were in arrears for the first three calls should be sued. Because of the state of the funds in 1795 and 1796, the company made a number of attempts to borrow money from banks in Edinburgh and Glasgow. In February 1796 Rennie gave up his previous half-year's salary because he had found it unnecessary to attend to the works as often as he had expected when he took the post; and this 'liberal offer' was rewarded with a piece of plate valued at 20 guineas.[9]

In March 1796, the report of the governor and directors gave several reasons for the slow progress of the work. These included the remoteness of the canal from the 'general places of business in Scotland', which had made many preparations necessary that would not otherwise have been required, and the difficulties encountered in procuring ground, particularly from John MacNeill at Oakfield. The actual construction had not begun until September 1794, and the completion of the canal, planned for the end of 1796, had been held up by the 'unfortunate situation of public affairs', which had made money scarce. An optimistic view was taken of the state of the works, despite the 'tardy manner' of execution. It was expected that the cutting between Ardrishaig and the summit level would be completed in the course of the summer; and it was said that work on the locks, aqueducts, culverts and drawbridges had 'proceeded with equal success'. It had been hoped that the canal would be completed in 1798, but in May of that year Rennie said that money was coming in so slowly that he could not fix a completion-date. It seemed likely, he said, that 'owing to bankruptcies' the capital would fall about £14,000 short of the amount subscribed.[10]

In his 1798 report Rennie gave a detailed description of the canal as it was then envisaged. The eastern end was to be 48 ft wide at bottom, 84 ft wide at surface and 13 ft deep. The summit level was to be 42 ft wide at bottom, 84 ft wide at surface, and 14 ft deep. At the west end there was rock, and the canal there was to be 30 ft wide at bottom, 56 ft wide at surface and 13 ft deep; but through the rock by the side of the bay of Crinan it was to be only 25 ft wide at bottom and 48 ft wide at surface. There were to be two towing-paths, one for westbound and one for eastbound vessels. The number of locks was to be 15, and the summit was to be 63 ft above sea-level. The locks were to be 96 ft long and 24 ft wide, except for the last two at the western end which were to be 112 ft long and 27 ft wide. These larger locks were to give admission to a dry-dock where coasting vessels too large to navigate the canal could be repaired. The rise of the ordinary tide in Loch Gilp was about 10 ft, and there would be 8 ft of water at the entrance lock at low tide. The rise in Loch Crinan was about 9 ft, and this would also leave 8 ft at low tide. Rennie had no doubt in 1798 that the canal would be used not only by busses and boats passing between the northern herring-fisheries and those in Loch Fyne but also—because of the war—by vessels returning from the Baltic to the ports of the English Channel; and it was expected that much of the Clyde's trade with the Baltic and the West Indies would pass through the Crinan.[11]

Efforts were made in 1799 to raise the £22,000 needed to complete the canal. Consideration was given in April to a proposal that the proprietors should subscribe a further £20 in respect of every £50 of the company's stock; and in May the company was given authority to borrow £30,000 by assignment of tolls, or by bonds, or by creating new shares. The Act pointed out that they had made great progress in carrying on the works, but that they could not complete them because several of the subscribers had been unable to answer the calls made on their subscriptions. Labour was also in short supply, and in May 1799 the governor was asked to apply to the Commander-in-Chief in Scotland for any soldiers who could be spared from military duty and would be willing to engage in the company's works. Wage increases resulting from war-time inflation had added to the cost of the canal, and the undertaking was in danger of being abandoned. The repayment at this time of the government loan of £50,000 by the Forth & Clyde was of importance, as a loan of £25,000 was then made by the Treasury to the Crinan on the security of the canal's rates and duties. A condition for the loan was that the canal should be assigned on a mortgage to the Barons of the Court of Exchequer in Scotland until the sum borrowed had been repaid with 5 per cent interest.[12]

Difficulties over food supplies arose in 1800, a year of bad harvest in Scotland; and in March it was decided to buy 100 quarters of oats or barley from London for the workmen. Work continued on the canal; but money remained scarce, and efforts were made to find purchasers for the remaining shares of the company's stock at the rate of £30 a share. A refusal was received however, from Drummond's the bankers, who said they made it a rule 'not to engage in any undertaking of this nature'.

A subscription to raise £9,000 was opened among the proprietors in March 1801, the purpose being to 'open the canal and to discharge debts'; and it was resolved that those in arrears who had paid half or more of the amount of their shares would forfeit them if they were not fully paid up by 1 May. By May, this subscription amounted to £9,810 and included £600 each from the Earl of Breadalbane, William Pulteney and D. Campbell, and £420 from the Earl of Breadalbane 'for the Easdale Slate Company'. Interest on this loan was paid 'for a year or two' only.[13]

The canal was opened in an unfinished state on 18 July 1801. Water was let in only by 'slow degrees', because it was thought that until the banks had 'settled and consolidated' the full depth might impose too much strain. In April 1802 there was 8 ft of water throughout the canal, and it was confidently hoped that when all the works were completed the undertaking would be 'attended with profit to the proprietors'. The number of shares which had been forfeited was 307, and there was doubt as to whether the 'great and useful work' could be perfected without further government aid. In March 1803 it was said that the canal's unfinished state was discouraging some vessel-owners from using it: the depth had still not reached 10 ft, the entrance at the west end was 'imperfect', and interruptions to trade were frequent. Yet, although the western fisheries had failed in 1802 and prejudices were thought to have operated at first against the 'new mode of navigation', tonnage dues for the canal's first 20 months amounted to £1,319, which was regarded as fairly satisfying. The canal had three reaches, the Ardrishaig or east reach, the summit reach, and the Crinan or west reach. The east reach was nearly 4 miles long and 32 ft above sea-level, and was connected with the sea at Ardrishaig by 4 locks and 3 basins. The summit reach was 1,114 yd long and 64 ft above sea-level, and was connected with the east reach by 4 locks and 3 basins at Cairnbaan and with the west reach by 5 locks and 4 basins at Dunardry. The west reach was about 3 miles long and 18 ft above sea-level, and was connected with the sea at Crinan by 2 locks and 1 basin. The dimensions of the locks were as Rennie had stated them in 1798; but the surface widths were of course less than he had then intended.[14]

Monthly reports by John Paterson, the company's resident engineer, disclose some of the problems involved in the construction of the canal. Skilled and unskilled labour was scarce: the contractors responsible for the work at the west end proved very unsatisfactory, and Paterson had to go south in 1797 to try to find new ones. This was not easy, since so much building was going on in the Lowlands that the 'most respectable tradesmen' were fully employed. Masons were needed to build the locks, but it was a hard task to attract them to Argyll. Carters also were given 'great encouragement' to come and work for the company, as it was 'absolutely impossible' to get carters in the district to execute the work. In July 1799 the workmen were making the most of the situation and becoming 'very unreasonable in their demands'; this boldness was attributed largely to the prospect of alternative employment offered by the herring-fishery. The masons were 'not less troublesome', although their wage of 16s a week was 4s more than that of the labourers. The workers had all been engaged for the whole season, but Paterson expected some of them to leave without notice. He saw a connection between high wages and discontent, and was convinced that it would be 'improper to listen to the demands' which were being made. Uncertainty about whether money to complete the canal would be available in 1799 resulted in a large-scale departure of the workmen, so that the few men who remained could ask for 'almost what wages they pleased'. Most of the labourers were local men, but those who were skilled workers, like the masons and the carpenters, came from the Lowlands. Later on, in 1806, there was competition for labour from the Caledonian; and in March of that year it was decided to make an advance in day-labourers' wages as a means of keeping 'most of the useful ones' who might have been attracted to the larger undertaking.[15]

The supply of stone for the canal came mainly from Arran, Morvern and Mull. There was often difficulty in transporting the stone because of a shortage of vessels: in February 1797, for instance, Paterson was sent to the Lowlands to 'engage so many vessels as could be had on reasonable terms to carry stones from Morvern to the sea-lock in Portree'. In May 1798 (probably as a result of this search) three sloops were employed in transporting freestone to Crinan; but the locks at the summit were built of rubble masonry instead of freestone, with the result that they were frequently in need of repair. Timber, especially oak, had to be sought in the Lowlands. In July 1795 Paterson undertook to purchase 383 cu ft of oak at 3s 4d per cu ft from James Lindsay of Glasgow, and about 3,000 cu ft of Danzig fir at 2s per cu ft from a company near Port Dundas. Most of the timber for the drawbridges and locks was American oak of inferior quality, which within a few years became 'totally rotten'; this was due in part to the wartime difficulty in obtaining timber from the Baltic and in part to the company's lack of money. (Rennie had not encouraged the use of American oak, as he had told the company's clerk in December 1793 that it was the 'worst of all oak' and 'would not last six years in the work'.) A supply of 500 to 1,000 tons of iron was applied for in February 1797 from the Earl of Eglinton's estate; prices for iron had risen as a result of wartime demands and an additional 3 guineas per ton for wrought iron had had to be paid earlier in 1795 to one of the company's blacksmiths.[16]

One of the chief obstacles in the canal's construction was the unexpected hardness of the whinstone which alternated with peat moss along the line. The moss near the western end defeated attempts to drive piles in, and this made it necessary to cut into the rock of the Knapdale Hills on the south. The embankment in the bay of Bellanoch took a long time to complete, and in December 1799 over 100 men were at work along the shore of Loch Crinan, where blasting was being carried on. The rocky projections which remained were very dangerous to vessels, and an 1803 report on the state of the rocks said that they were generally hard and difficult to cut, their projections being 'extremely bold' and the cavities large. It would have been expensive to fill the cavities with cement, and the projections could not be cut off without pieces of rock falling into the canal; so it was decided to put up wooden fenders.[17]

The canal was still incomplete in 1804, and a meeting was held to discuss an application for government aid to enable the company to complete the work. It was claimed that £140,610 had been expended on the canal, over £108,000 of this having been raised by private subscription, and that £25,000 was needed to complete the navigation and pay the contractors' and tradesmen's debt. This application was successful, and an Act was passed authorizing the Treasury to advance £25,000; the conditions were the same as those for the previous government loan. On 7 January 1805 about 70 yd of the canal banks in 'Oakfield moss' were destroyed by floods, and a new line had to be dug. In May 1806 a pier was made at Ardrishaig with stone from Arran; and in July the same year the canal was reopened to traffic with a depth of 'nearly 9 ft of water'. James Hollinsworth, the resident engineer whom Rennie had recommended in 1805, estimated the cost of finishing the canal at £5,992; but this included the construction of a dry-dock at Crinan, and the company decided that in view of the state of the funds this would have to be dispensed with. In 1808, however, work was started on 'new reservoirs', and in February the following year these were completed. The canal was pronounced 'finally complete' in August 1809, and a programme of retrenchment, including some reduction of salaries, was imposed to prevent the company's debt from increasing.[18]

Unfortunately for the company there was little hope of economy, as the canal had been built so badly that heavy expenditure was often necessary for repairs. In January 1811, after a violent gale from the south-west, the embankment of the principal reservoir at Glen Clachaig collapsed and water rushed into the canal, sweeping away part of the road at Lock 5 and dismantling the lower gates of Lock 6. The canal was closed to navigation, and between 50 and 60 labourers were employed to repair the canal-banks and clear the cut. In May 1811 Lord John Campbell presented a memorial to the Treasury asking for a loan of £5,000 on the same terms as previous loans, and this sum was granted out of the Consolidated Fund. Although the canal was partly repaired, however, and reopened in January 1812, money was still urgently needed. Another memorial presented in June 1812 said that all the funds of the company were exhausted and that more money was wanted to repair the reservoirs and undertake 'other necessary work'. The memorial went on to say that there was no hope of raising money privately, and that unless public aid was made available the canal would have to be abandoned. It was therefore requested that some 'eminent engineer' should be employed by the government to make a survey of the state of the canal and estimate the cost of completing the work. The Treasury then wrote to the Commissioners of the Caledonian asking that Telford should make this survey.[19]

At this time Isaac Holder was resident engineer of the Crinan,

James Hollinsworth (who had succeeded his father in January 1811) having been dismissed for incompetence in January 1812. Describing how the collapse of the Crinan drawbridge had caused the death of the bridge-keeper's daughter, Holder stressed the need for immediate repairs to all the bridges; and some temporary but unsatisfactory repairs were made. The canal's water-supply was very precarious, and Holder feared in January 1813 that there would not be enough water in the summer to keep the canal navigable. The gates at Locks 14 and 15 were in a decayed state and were said to be infested with sea-worms; but the repair of the reservoir at Glen Clachaig remained the most urgent need.[20]

In January 1813 Telford made his report, saying that the canal was in 'a very imperfect condition'. At the eastern entrance there was a range of rocks about 500 ft long which was only visible at low water and made the entrance and departure of vessels very dangerous. The ground outside the sea-lock at the western end was about 4 ft higher than the top of the lock-sills and prevented vessels of any considerable draught from entering or leaving except at high tide. There were leaks in the masonry of some of the locks, and some of the lock-gates were decayed. All the drawbridges were in a bad state of repair, and the rock-cutting near the west end had left the sides so rugged as to injure the sides of the vessels that struck against them. The damhead of the principal reservoir having been carried away, the canal was impassable in dry weather for want of water. Telford's estimate for the necessary repairs was £18,251; and as £55,000 had already been borrowed from the government, the company's financial position was now very serious. Telford had no doubts about the utility of the canal, and considered that the 'intercourse of that part of the kingdom would be materially injured if this communication was now abandoned'. On the future of the canal, he said that it could 'never be any object to the original subscribers', and that if it was decided to proceed with the work the 'most advisable scheme' would be to put it under the direction of the Commissioners of the Caledonian.[21]

Hugh Baird, the engineer associated with the Forth & Clyde and the Union, made a report in October 1814 on the state of the Crinan reservoirs. He named the reservoirs as Glen Clachaig, Camloch & Donloch, Loch Clachaig, Loch na Faoilinn and Loch-an-Add, and suggested the raising of their damheads in order to increase the supply of water. He could not account for the collapse of the Glen Clachaig damhead, but took the view that it was the kind of accident that could have happened to 'the best damhead that could be made'. By April 1815 the damheads of the Loch Clachaig and Camloch & Donloch reservoirs had been raised 5 and 5½ ft respectively; but despite these improvements the state of the canal was becoming worse. The shallowness of the entrance at Ardrishaig was a serious problem: in October 1815 a sloop loaded with herrings, being detained because of the lack of water, was towed out of the harbour to a ridge of rocks at the head of the breakwater and was there wrecked by a storm. In February 1816 the canal was said to be in 'good working order'; but this was qualified by a list of exceptions, chief of which were the lock-gates at Crinan, and in May the western end of Bellanoch Bay was discovered to have numerous leaks caused by the original careless execution of the puddling on the banks.[22]

The revenue from tolls in 1802 was only £810, but this rose by 1810 to £1,178; and the number of passages made went up from 668 in 1804 to 1,578 in 1810. The canal was reopened after repairs in 1812 and kept open till 1816, 'notwithstanding the incomplete and decayed state of the works'; but revenue fell because of the lack of public confidence. One of the company's bankers, Carrick Brown & Company of Glasgow, made an urgent application in April 1816 for the payment of money advanced in May 1808 for the construction of reservoirs; but the company was unable to pay this sum. In April 1816 a petition was sent to the House of Commons asking for the canal to be put under the direction of the Commissioners of the Caledonian and completed at the public expense. The petitioners stressed the canal's future importance as a complement to the Caledonian in providing a communication from the Clyde to Inverness. Carrick Brown were asked in June to postpone their demand, as the government would not advance any money to pay the company's debts, and later in the same month an Act was passed which authorized the Barons of the Exchequer to issue £19,400 to the Commissioners of the Caledonian Canal for the completion of the works on the Crinan. The canal was to remain vested in the Barons of the Exchequer until all debts were paid: this meant that this loan and the three earlier ones with interest at 5 per cent had to be repaid before the company could resume control.[23]

The funds which had so far been expended on the construction and maintenance of the canal consisted, according to the evidence later given before a Select Committee of the House of Commons, of £108,146 raised by private subscription, £74,400 borrowed from the government, and £2,000 borrowed privately. Of the original subscriptions of £50, 1,851 had been paid in full, so that the company's stock was £92,550; and another £5,786 had been paid in interest by subscribers whose payments had come in late. The additional subscription of 1801 had brought in a further £9,810.[24]

Telford was responsible for the direction of repairs to the canal; in December 1816 he was employed in making a detailed inspection of the state of the works, and by January he had procured a contractor to undertake the principal repairs. These included extending the pier at Ardrishaig, deepening the eastern entrance, replacing the drawbridges with cast-iron ones, repairing or renewing all the lock-gates, repairing the bottom of the Crinan sea-lock (for which a steam-engine was necessary), repairing all the lock-bottoms, and widening and straightening some parts of the canal amongst the rocks at the western end. The canal was closed on 1 March, and in that month the repairs were begun; but stormy weather interrupted the work and delayed vessels bringing materials from the Clyde. John Gibb, who had directed the Aberdeen harbour works, was responsible for most of the repairs. Rails, tools, a steam-engine and a cargo of pitch-pine lock-gates were safely delivered; and carpenters from the Lowlands were employed on framing the gates. A large number of labourers worked to remove a part of the canal-bank known to the workmen as 'New York Bay' and to replace it by a straight line.[25]

Further supplies of timber arrived from Greenock and Liverpool in April 1817; and in that month there were 15 masons, 20 carpenters, 9 smiths, and 216 labourers at work on the canal. The improvements between Dunardry and Crinan continued in May, the main task being the straightening of a number of bends within a short distance of each other. The work was not completed without accident; in August a labourer fell into the lock at Crinan and was drowned, leaving a poor widow and three young children; and Telford, on behalf of the Commissioners, authorized the payment of £10 as compensation. By September the work was being brought to a close, and half the labourers were discharged. Telford inspected the work in November and declared himself satisfied with the state of the improvements and repairs; and on 20 November the canal was reopened. It was now under the management of the Commissioners of the Caledonian; but its profits were at the disposal of the Barons of the Exchequer. The salaries and wages included £220 for the resident engineer, £75 for the Secretary of the Board of Directors in London, £41 for the carpenter, £36 each and free houses for the 3 lock-keepers, £35 for the collector of toll dues, £31 for the foreman in charge of the day-labourers, £10 for the man in charge of the pier headlight, and £5 each and free houses for the 3 bridge-keepers. Trade improved slightly after the reopening: toll revenue in 1818 was £1,640 compared with £1,248 in 1816, and in April 1819 traffic was said to be greater than in the corresponding month of any previous year, thanks to the increased demand for slates from Easdale. The number of passages increased from 1,696 in 1818 to 3,028 in 1819, Henry Bell's steamship the *Comet* having begun to ply a profitable trade between Glasgow and Fort William; but on 13 December 1820 the *Comet* was wrecked near Craignish Point. The owners of its successor, the *Highland Chieftain*, applied for the privilege of free transit which had been granted to the *Comet*; and William Thomson (who had been resident engineer since 1814) accorded this for a limited period. The *Highland Chieftain* was replaced in 1822 by *Comet II* and the *Highlander*, which together maintained a twice-weekly service.[26]

The revenue from the herring trade fell from £567 and £486 in the years 1818 and 1819 to £209 and £192 in the years 1822 and 1823; and slates yielded £378 in 1819 but only £189 in 1823. The kelp industry, which had been expanding in the early years of the century, declined after the halving of the duty on barilla in 1822; revenue from kelp in the year 1823 was only £40. Revenue from the steamboats on the other hand, rose from £187 in 1821 to £286 in 1823; and the number of passengers conveyed by them rose from 2,400 in 1820 to 6,939 in 1823. As the result of an 'urgent request' from the owners of the *Highlander*, coal-stores had been provided at each end of the canal; but in May 1822 Thomson pointed out that since the advent of steamboats 'strangers' had frequently criticized the canal authorities for not providing a 'comfortable house' for travellers. The Caledonian Canal opened in October 1822, and in December 1823 plans were announced for a steamboat service through the two canals from Inverness to Glasgow; the journey was to take 3 days, with overnight stops at Fort William and Crinan. In 1824 it was established that steamboats exclusively for the conveyance of passengers were to be charged according to their register tonnage, being allowed 56 lb of luggage for each passenger but paying 1s for every package above that allowance. By 1825 the Glasgow–Inverness route was being served by two operators, one with *Comet II* and the *Highlander* and the other with the old *Highland Chieftain* and the *Ben Nevis*. The introduction of these paddle steamboats proved detrimental to the banks, and especially to part of the bank north of

XIII. Aberdeenshire Canal: (*above*) a passage-boat near Kittybrewster about 1825; (*below*) the abandoned canal near Pitmedden about 1930

XIV. Crinan Canal: (*above*) the sea-lock at Ardrishaig; (*below*) Queen Victoria's passage through the canal in 1847

Dunardry, which was composed of moss. Most of the vessels which used the canal were under 60 tons burden. As an inducement to larger ships to use the canal it was determined in June 1824 that all vessels exceeding 60 tons were to be charged dues for 60 tons only. For the regular traders taking cargoes to and from Glasgow a depth of 9 ft was sufficient; but it was possible to keep the depth of water at 10 to 11 ft. Dues were 1½d per ton per mile on coal, lime, manure and salt outward bound, and 2d per ton per mile on all other goods. The expense of trackage was generally 6s or 7s on a vessel of average size.[27]

The Commissioners had decided in 1820 that expenditure on the canal should be kept down, and that no loans could 'at present' be repaid. Revenue for the years 1818–35 averaged about £1,720 per annum. A letter sent to the Treasury in 1825 stated the canal's precarious situation: revenue had not come up to expectations, and instead of being over £3,000 per annum as Hugh Baird had estimated, it had fluctuated in the years 1818–24 between £1,371 and £2,031. The surplus revenue had been used to keep down the interest due on the canal's debt; but this interest had none the less risen by 1820 to £3,000. In 1826 fears had been expressed that the revenue was liable to become insufficient for the maintenance of the canal, since the demand for repairs was increasing. In particular, cast-iron tunnels were needed for some of the reservoirs, and these extra repairs presented the prospect of expense beyond the funds in the possession of the Commissioners, which were at that date no more than £918. The number of people passing through the canal in steamboats increased from 8,332 in 1824 to 13,824 in 1827. The facilities offered by these boats were limited by their small size, a consequence of the dimensions of the canal. The want of suitable accommodation was a 'constant theme of complaint', and so was the boats' inferior power and safety compared with those of vessels rounding the Mull of Kintyre. Rivalry among the steamboats was disliked by the canal authorities because of the risk of damage to the banks, and in May 1829 the owners were told that the boats would not be allowed to pass through the canal together.[28]

In May 1834 a suggestion that the canal should increase its rates and duties was turned down because (as Thomson wrote to David Caldwell, secretary both of the Crinan Canal Company and of the Forth & Clyde Canal Company) the Mull of Kintyre provided an alternative route which 'from increased experience and better vessels' was 'less an object of dread than formerly'. Thomson said that a steamer of 120 hp was being built at Greenock for the Clyde and West Highland trade round the Mull. It was now evident that the development of steamboats had destroyed all possibility of the existing canal becoming profitable; but Thomson felt that if the cut were larger 'this opposition to the Crinan Canal interests would not be attempted'.[29]

The condition of the canal was deteriorating rapidly; and in May 1835 a long list of 'necessary repairs' was drawn up. A contract for the most important and pressing repairs was made in June 1835 with John Gibb of Aberdeen; the tasks undertaken included the raising by 5 ft of the reservoir at Loch-an-Add. In August the canal was forced to close, part of the embankment between Locks 11 and 12 having broken because of unprecedented water-pressure; and in September James Loch, one of the Commissioners of the Caledonian, made a full report on the works. He said that no vessel could lie along the breakwater at Ardrishaig because of its shelving construction, and that passengers were therefore forced at low tide to 'slide down in a very inconvenient and somewhat dangerous manner'. The presence of steamboats by the pier and of herring-boats in the harbour impeded the entrance and departure of other vessels, and Loch proposed that the breakwater should be extended and that two new piers should be constructed on the opposite side of the bay. The lower gate of Lock 4 was held together only by iron clamps; and the defects in the western section, where the canal had been cut out of solid rock, were such as could not be remedied at any 'reasonable' cost. The canal's locks admitted only small vessels which were unsuited to the exposed voyage from Crinan to Corpach; and the number of steamboats going round the Mull was increasing. It was hoped that the Caledonian, when completed, would admit boats 40 ft wide; and Loch suggested that goods and passengers might be transhipped at Crinan. Whether the construction of the Caledonian and the Crinan had been a 'wise outlay' or not, Loch concluded, these canals had brought great advantages to the 'most remote and sequestered districts of the West Highlands and Islands of Scotland' by 'enabling the poorer classes of the people to convey their eggs and poultry and summer labour to Glasgow and its vicinity, and to obtain their supplies in return at a cheaper rate and in greater plenty'; and this had in some measure compensated for 'the loss of the kelp manufacture'.[30]

A new pier north of the canal entrance at Ardrishaig was completed in March 1837. In the same year James Walker, who had been commissioned by the Treasury to report on the state of the Caledonian, was invited to inspect the Crinan as well. His report appeared in July 1838, and suggested the extending of the Ardrishaig breakwater, the deepening of the eastern entrance by at least 5 ft, the lowering of the summit level, the widening of those parts of the western reach where it was impossible for vessels to pass, and the replacement of the lock-gates at Crinan. The eastern entrance was deepened by 2½ ft in 1838, and the number of passengers carried by the steamboats was reported to have risen in that year to 21,406. Two new boats, the *Inverness* and the *Helen McGregor*, had come into operation in 1835; and in 1839 Burns, Thomson & McConnell instituted a co-ordinated service between Glasgow and the West Highlands, with the new steamer *Brenda* working between Glasgow and Lochgilphead, the new track-boat *Thornwood* working on the canal, and the old steamers providing the connections from Crinan.[31]

James Walker and David Caldwell were the chief witnesses before the select committee on the Caledonian and Crinan Canals whose report was published in August 1839. Walker reported complaints about the poor accommodation at Ardrishaig, the narrowness of the western reach, and the fact that the canal was shut at night. Caldwell said that the original company had tried 'on several occasions' to regain possession of the canal, and that the decayed state of the works and the high level of the dues were still the subject of frequent complaints from the shipping trade. Neil Malcolm, a local landowner who had inherited canal shares, mentioned complaints that the resident engineer, William Thomson, gave too little of his time to the canal, and explained that Thomson was a sheep farmer and a wool merchant and had connections with a timber yard at Ardrishaig. In his evidence, Thomson claimed that the canal could be navigated by steamboats in 3½ hours and by other boats in about 4 hours. He gave the revenue as £1,700 to £2,000 per annum, the dues on cargoes as 1d to 3d per ton per mile, the tonnage dues on steamboats as 1d per ton per mile, and the wages of day-labourers as 1s 6d to 1s 8d. Most boats using the canal, he said, were engaged either in the West Highland herring-fishery or in the transporting of goods between the Clyde and the islands; but many cargo-vessels went round the Mull of Kintyre. The canal could accommodate vessels of 160 tons burden, but the typical coaster was of 20 to 60 tons. The committee's report emphasized the canal's dilapidated state, navigational hazards and inadequate depth. Despite the benefits the canal had brought to Argyll and the north-west, the committee recommended that the Treasury should be empowered to make arrangements for its future management before any further outlay was risked.[32]

Steamboat revenue for 1839 was £322 below the average for 1833–7, partly because of competition from boats going round the Mull and partly because many steerage passengers were deciding 'from motives of economy, to walk the length of the canal'. By March 1840 the gates of the Crinan sea-lock had been repaired and sheathed in iron, the reservoir at Loch Clachaig had been fitted with an iron tunnel and sluice, and the entrance at Ardrishaig was being deepened. In July the resident engineer received letters of apology from a number of canal employees, who acknowledged their 'misconduct in tippling at Ardrishaig'; fines of 2s 6d and 10s 0d were imposed, and one of the lock-keepers was discharged and told to vacate his house immediately. In February 1841, Thomson wrote to the Rev David McLachland, sending 45s for the poor of Lochgilphead; the money, he explained, had come from fines imposed 'as a check on the improper use of whisky occasionally indulged in by some of the canal workmen, and particularly on 15 July last'.[33]

The revenue for the year 1839–40 was £2,455, the highest figure to date; the increase was attributed to the prosperity of the north-west herring-fishery. Traffic at Ardrishaig in 1841–2 included 13,398 passengers landed, 11,484 passengers embarked, 15,410 sheep and lambs, 832 pigs, 415 cattle, 1,851 boxes and barrels of herrings, 1,196 tons of coal, 1,226 tons of general goods, 41,271 register tonnage of steamboats and 1,557 register tonnage of sailing-vessels. A new pier was erected at Crinan in 1843 for the benefit of the steamboats which called there. From 1840 onwards, a regular passenger-service between Ardrishaig and Crinan was maintained during the summer months by two iron vessels pulled at 8 to 10 mph by thoroughbred horses with riders in scarlet livery; but it was said in 1846 that these boats were used by only 2,000 passengers per annum. In 1830 Henry Bell pointed out that the invention of steamboats had removed some of the objections to the Tarbert Canal scheme; and in 1846 an Act was passed to authorize the making of a canal 'from the harbour of East Tarbert on Loch Fyne to join West Loch Tarbert at or near the north-east end'. The subscribers, who included Walter Fred Campbell of Islay and Archibald Campbell of Jura, formed an Argyll Canal Company; but the raising of capital proved impossible, and the company was dissolved in 1849, its powers to buy land having then expired. The Crinan was closed because of drought in December 1844 and January 1845; and in 1846 the reservoirs of Loch na Faoilinn and Glen Clachaig were united in an attempt to increase the water supply.[34]

In June 1844 the Commissioners of the Caledonian Canal expressed their dissatisfaction with the 'very undefined state of their connection with and powers over the Crinan', and asked their secretary to bring the matter to the notice of the Treasury. The original company had made several attempts to regain possession of the canal, but had been unable to raise the money to pay off their debts. The Commissioners controlled the canal's wharfage and tonnage dues and administered its revenue; and they had incurred a new debt of £1,500 in order to build a pier at Ardrishaig. In November 1844 the Treasury replied that the canal was to be transferred from the Barons of the Exchequer to the Queen's Remembrancer, and asked the Commissioners if they would be willing to undertake the duties of management. The Commissioners agreed to this, but suggested that the transfer be delayed until a legal dispute about the rights of a nearby landowner had been settled. In August 1848 an Act was passed vesting the Crinan Canal in the Commissioners of the Caledonian. The canal was to revert to the original proprietors when they had repaid their public debt of £74,400 with interest and refunded such sums as the Commissioners had expended, in excess of the canal's toll receipts, on the repair and improvement of the works; this situation, however, was never reached.[35]

One of the highlights in the Crinan's history was Queen Victoria's passage through the canal in the track-boat *Sunbeam* in August 1847. From Ardrishaig, Thomson wrote to his son in London that there was 'a greater stir than ever you saw in this little place' and that he had made arrangements for the 'speedy passing of the royal barge along the canal and through the locks'. Queen Victoria wrote in her journal: 'We and our people drove through the little village to the Crinan Canal, where we entered a most magnificently decorated barge, drawn by three horses ridden by postilions in scarlet. We glided along very smoothly, and the views of the hills—the range of Cruachan—were very fine indeed; but the eleven locks we had to go through were tedious, and instead of the passage lasting one hour and a half it lasted upwards of two hours and a half.' The return journey on 18 September was speedier; the queen recorded that 'there was a piper at each lock', but that 'it rained almost the whole time'. At Ardrishaig, according to *The Times*, 'Her Majesty, Prince Albert and the royal children entered a closed carriage and drove down at a walking pace to the quay, off which lay the *Black Eagle* and *Undine*. A little before the embarkation rain commenced. Her Majesty appeared somewhat chilled. Prince Albert shielded Her Majesty from the pelting rain by holding an umbrella over the royal head.' One result of the visit was an increase in the canal's tourist trade.[36]

In the autumn of 1847, Charles Herbert succeeded Thomson as resident engineer. In the following year he reported that some of the lock-gates needed to be renewed, and that the lower sill of the Crinan sea-lock was in need of repair. In 1850 the upper gates of Locks 8 and 14 were renewed, and the harbour-entrance at Ardrishaig was deepened nearly 2 ft with the help of a small dredging-machine. In 1851 there was an increase in steamboat passages; but it was said that most of the herring caught in the west-coast fishery were now being shipped round the Mull. Writing to Samuel Smith in the same year, Herbert reported that a small stone wall had been built on each side of the towpath at Cairnbaan; the need for this wall had become evident on an unusually dark night when five people, four of them perfectly sober, had fallen into the canal at the same spot.[37]

By 1850 most of the West Highland trade was in the hands of G. & J. Burns; and in 1851 this firm was taken over by its chief clerk, David Hutcheson, who formed a partnership with Alexander Hutcheson and David MacBrayne. The passenger-traffic in the 1850s was carried by the steamer *Pioneer* between Glasgow and Ardrishaig, by the track-boats *Sunbeam* and *Maid of Perth* on the canal, and by the steamer *Shandon* between Crinan and Oban, where there were connections for Tobermory and Fort William. During Glasgow Fair Week, Hutcheson's advertised cheap excursions to Staffa and Iona 'in order to afford the operative classes an opportunity of viewing the islands'; the fares from Glasgow Bridge were 10s for a single person and 12s 6d for a married couple. Three other steamers—the *Cygnet*, the *Lapwing* and the *Duntroon Castle*—were trading between Glasgow and Inverness in 1851; and by 1854 the last of these had been replaced by the *Maid of Lorn*, later renamed the *Plover*. The introduction of larger ships on the service between Glasgow and Loch Fyne soon created a demand for better accommodation at Ardrishaig, where a 50-ft quay had to serve such vessels as the *Iona*, which was 235 ft long. Rising food prices having forced a general increase of wages in the district, Herbert raised those of the canal labourers to 1s 10d or 2s 0d per day and those of the carpenters to 3s per day. The canal's traffic in 1854–5 included 33,280 passengers, 27,200 sheep, 2,080 cattle and horses, 4,230 boxes and barrels of fish and 2,450 tons of goods. In 1857, when John Fyfe replaced Herbert as resident engineer, the canal revenue was adversely affected by a decline in the slate trade from Easdale and Ballachulish. A report to the Treasury in the same year, however, said that trade as a whole had doubled since 1842 and that merchants had declared themselves willing to pay increased tolls in return for better facilities. In August 1857 an Act was passed authorizing the erection of such piers and jetties as had become necessary to deal with the increased number of passengers and animals being carried. This work was to be paid for by new pier dues of 2d each on passengers, 6d per score on sheep, goats and pigs, 6d each on cows, 1s od each on horses, 6d each on empty and 1s od each on loaded carts, 2d per ton on coal, 3d per 1,000 on slates, bricks and tiles, 2d per bag on wool and 1d per 50-ft load on timber.[38]

The canal's trading prospects were described in 1858 as 'encouraging', partly because of its value to the agriculture of 'an improving district' and partly because of its importance in linking the Caledonian with the Clyde. Most of the passengers arriving at Ardrishaig went through the canal in track-boats and proceeded northwards to Corpach; and much of the canal's goods traffic followed the same route. The number of passengers who had embarked or landed at Ardrishaig in the preceding year was given as 44,537, and the total tonnage of the steamers which had visited the harbour there as 56,566. In August 1858 the Commissioners introduced new rates of 1s per ton on unladen sailing-vessels, 1s 3d per ton up to 100 tons and 1s per ton thereafter on sailing-vessels carrying coal, slates, bricks, kelp, herring and manure, 1s 6d per ton up to 100 tons and 1s per ton thereafter on sailing-vessels carrying other goods and on pleasure-yachts, 1s 9d per ton on vessels fitted with screw-propellers, and 2s per ton on steamboats.[39]

In February 1859 the Camloch embankment gave way, and the water descended through the lower reservoirs of Loch Clachaig and Glen Clachaig into the canal, filling over a mile of it with debris. George May, the resident engineer of the Caledonian, inspected the damage and put the cost of repairing it at £10,000. At the same time, he suggested a number of ways in which the canal could be improved, and raised the possibility of replacing it either by a new ship-canal or by a railway. The Commissioners rejected the railway and ship-canal schemes, but recommended that the summit should be lowered and the locks lengthened. The Treasury in its turn rejected these proposals, but authorized the expenditure of £12,000 on repair-work. James Walker, who had estimated the cost of the Commissioners' scheme at £80,000, was sent to Ardrishaig to supervise such reconstruction as was possible with this smaller sum. By December 1859 the repairs account was overdrawn; and in January 1860 the Treasury granted another £2,000 to finish the work. Seeing that the canal was likely to become a permanent burden on the public, the Commissioners endeavoured in February 1860 to arrange a 99-year lease of the work to a private concern. This plan was given legal sanction in July, but no satisfactory offers were received. The canal reopened in May 1860, but work on Loch Clachaig went on until 1864.[40]

The introduction of screw-steamers on regular services in the 1860s had an adverse effect on the canal's revenue, since the tonnage dues for these vessels were lower than those for paddle-steamers; but the Commissioners remedied the situation by establishing the same dues for both types. In 1866 Hutcheson's replaced their track-boats by a twin-screw steamer, the *Linnet*, hoping that the improved accommodation would attract more traffic. Another screw-steamer, the *Chevalier*, was introduced on the service out of Crinan; and a new steamer service was established between Ardrishaig and the North British Railway station at Helensburgh. In 1868–9, the year in which William Rhodes became superintendent of the canal, the revenue rose to £4,315; but it was still exceeded by the expenditure.[41]

In 1875 the Commissioners received a petition about the rates charged at Ardrishaig and the arbitrary way in which they were calculated. An inquiry was held in October before a committee including the Marquis of Lorne; and evidence was given about the poor condition of the quay and about alleged favouritism. It was stated that Hutcheson's luggage-van was charged only 1s whatever its load, whereas 'merchants' had to pay 3s to 5s according to the number of parcels in their carts. Complaints were also made about the jutting rocks in the western section, the high dues on the canal, the inadequate wharfage at Ardrishaig, Miller's Bridge, Oakfield, Cairnbaan, Bellanoch and Crinan, and the decayed sills of the sea-locks at both entrances. On the committee's recommendation the Commissioners decided in March 1877 to draw up a new table of dues, and this came into operation on 1 May the following year. In 1879 Hutcheson's introduced a new steamer, the *Columba*, which was over 300 ft long and was described as 'the finest river-steamer in Europe'; in conjunction with the *Iona*, it maintained a daily service between the Clyde and Ardrishaig, where a 127-ft extension to the breakwater had just been completed. The herring-fishery, however, which had been an important source of revenue for the canal, was now said to be 'entirely lost'; and it was expected that the new railway from Glasgow to Oban would divert much of the through trade between the Clyde and the north-west. The canal suffered another blow in 1880, when traffic was seriously interrupted by drought in the summer and ice in the winter.[42]

The paddle-steamer *Cygnet* was wrecked in the autumn of 1882 and was not replaced; paddle-steamers were now being ousted by screw-steamers like the *Rockabill*, chartered by Hutcheson's successor David MacBrayne in 1883. The Argyll Canal Act, which was passed in this year, authorized the raising of £200,000 to build a one-lock canal from East Loch Tarbert to West Loch Tarbert; but like the 1846 Act it remained a dead letter because no funds were available. New piers were completed at Crinan in 1881 and at Ardrishaig in 1884. In the latter year the tradesmen of Lochgilphead complained about the rates and dues levied at Ardrishaig, and some reductions were made; the new rates included ½d per stone on butter and cheese, ½d per cwt on iron, 1½d per cwt on wool and 1d per cwt on zinc.[43]

John Davidson, who had succeeded George May as superintendent in 1868, took charge of the Caledonian in 1885; and his place was taken by L. John Groves, who had earlier been responsible for reconstructing some of the Crinan's bridges. In 1887, in a letter to one of the Commissioners, Groves set out the case for a lowering of the summit level. This would result, he maintained, in speedier transit through the canal and a reduction in working expenses, whereas if improvements were not made the canal would lose all its passenger-traffic to the Clyde, Ardrishaig & Crinan Railway which Parliament had just authorized. These suggestions, like earlier ones of the same kind, were rejected on grounds of expense; and in 1892 an Act was passed authorizing the abandonment of the railway, for which no capital had been subscribed. The year 1890 saw the introduction of new rates on steam yachts and launches of 1s per register ton for through passages and 1½d per ton per mile for shorter distances, with half dues above 40 tons and maximum charges of 25s for open boats and 32s for those with cabin accommodation. In the same year the wooden bridge at Crinan Ferry was replaced by an iron one; and in 1892 a new swing bridge was erected near Bellanoch. By 1905 it was agreed that the Ardrishaig pier should be extended; the cost was estimated at £3,340.[44]

The 1906 inquiry was told that the Crinan's revenue averaged £6,000 per annum. In 1904–5, it had included £710 from passenger-steamers and from transit duty on 24,512 through passengers, £275 from 225 yachts, most of them sailing yachts, £32 from 102 fishing-boats, and £315 from pier tolls for 75,720 local passengers; but 79 per cent of the revenue came from the canal's goods traffic. A new item in this traffic was the whale-oil being brought through the canal from a recently-established factory in Harris; and the other goods carried eastwards included slates, stone, sand, timber, kelp and diatomite, the last of these being sent from Skye and used to cover pipes and boilers. The canal was used by about 100 small steamers; but others went round the Mull to avoid the delays caused by tidal conditions at Ardrishaig.[45]

Groves, in his evidence to the Commission, advocated the construction of a two-lock canal 20 ft deep, which would admit the largest steamers engaged in the West Highland trade. He was confident that this improved canal would be used by the larger fishing-boats and steam-yachts which had so far been forced to go round the Mull; and he also believed that it would bring a sharp increase in passenger-traffic by making it practicable to operate a through service between Oban and the Clyde. In existing circumstances, the Crinan-Ardrishaig steamer could accommodate only 270 passengers, whereas the Oban–Crinan and Ardrishaig–Glasgow steamers could take far more; passengers' luggage had to be taken from Crinan to Ardrishaig by road, and the journey through the canal took 2¼ to 2½ hours. The canal was in any case in a poor condition, many of the lock-gates having decayed and the embankment at the west end being defective; and the repairs to these things would cost £15,000. Such trade as the canal had, Groves explained, consisted largely of goods which were being shipped in small steamers in order to pass through the Forth & Clyde: the main items were coal going west and slates going east.[46]

Another witness before the commission was David MacBrayne, the chairman and managing director of the company which controlled most of the goods and all the passenger trade on the canal. At this date, MacBrayne's had two cargo-boats (the *Handa* and *Brenda II*) which maintained a regular service through the canal, and a saloon steamer (the *Linnet*) which sailed from Ardrishaig to Crinan and back twice daily from Whit Sunday to the end of September. In addition, they had formed the practice of chartering the canal's ice-breaker, the *Conway*, in order to cater for the peak tourist-trade during the Glasgow Fair. Their other vessels were unable to navigate the canal; but they opposed Groves' plan for enlarging it, fearing the effect of a three or four years' closure on the profitable holiday-traffic they had built up.[47]

The Town Clerk of Lochgilphead, on the other hand, supported the scheme, stressing the inadequacy of the existing canal for the shipping of livestock, and arguing that a ship-canal would bring benefits to a large area. Support came also from a representative of the Argyll County Council, who said the old canal had 'outlived its usefulness' and was 'practically worn out', and from the MP for Kilmarnock, who described it as 'quite obsolete' and 'hardly fit to be called a canal at all'. Two possible routes for the proposed ship-canal were described by Groves and by the Glasgow engineers Crouch and Hogg. One of these went further north than the existing canal and followed the River Add for a considerable distance; but the recommended route was a direct one, with locks at Badden and Dunardry and an approach channel in Loch Crinan. The proposed width was 80 ft on straight sections and 90 ft on curves; and the locks were to be 350 ft long and 59 ft wide. At the side of each lock, however, there was to be a smaller one 210 ft long and 30 ft wide, which was to be used by smaller vessels in order to conserve water. The cost of this scheme was estimated at £745,000; and there was also a plan, adding £51,000 to the cost, for the elimination of the Dunardry locks, so that a ship using the cut would have to pass through one lock only.[48]

The Secretary of the Congested Districts Board, however, pointed out that a ship-canal could be constructed at Tarbert for one-third of the cost of the Crinan scheme; and this proposal was backed by an Edinburgh civil engineer, who put the cost of a Tarbert canal 21 ft deep at £301,000. Consideration was also given to the possibility of building a ship-railway at Tarbert; but this received little support. The Commissioners, after examining these proposals, concluded that the traffic likely to use a new canal either at Crinan or at Tarbert would not be sufficient to justify the capital outlay. Groves, however, continued to press for a ship-canal, declaring that the smallness of the existing canal debarred modern shipping from using the 'most direct and most sheltered route between the Clyde and the West Highlands'; and in July 1913 the Convener of the Argyll County Council wrote to the Commissioners of the Caledonian to protest against the Government's 'disgraceful' behaviour in allowing their property to 'become a century out of date'.[49]

During the war, many of the canal's cargo-boats were requisitioned by the Admiralty and as no dues were charged on Admiralty vessels, revenue dropped to about £4,000 per annum. The Treasury granted a £500 loan in 1916; but in 1917 the Commissioners reported that extensive repairs would have to be carried out at once if the canal was to be maintained at all. The Treasury lent a further £1,300; but wartime inflation raised the cost of repairs to over £5,000. By an Act passed in 1919 the Caledonian and

Crinan Canals were transferred to the Ministry of Transport. In their last report, published in 1920, the Commissioners noted a 'partial resumption' of the passenger-service on the Crinan; but the canal's expenditure still exceeded its income by more than £2,000.[50]

A committee organized by the Ministry of Transport reported that tonnage had declined from 90,000 in 1905 to 75,000 in 1921. The coal trade had remained steady at about 40,000 tons annually; but the quantity of slates carried had dropped from 11,997 tons to around 300 because of foreign competition. The only increases were in the timber trade, which had been stimulated by the war, in the trade to and from the aluminium works at Kinlochleven and Foyers, and in the shipping of barley to the distilleries in Islay. *Brenda II* was still using the canal regularly, though the *Handa* had been abandoned; and though most of the passenger-traffic for the north-west went by rail to Oban, Mallaig and Kyle, the *Linnet* was still operating during the tourist season as a link in the steamer-service between Oban and the Clyde. MacBrayne's representative opposed plans for reconstructing the canal; but the spokesman of Glasgow Corporation recommended that the canal should be enlarged so that it could deal more efficiently with the trade in live-stock and wool. All witnesses were in agreement about the canal's importance to the people of the Hebrides; and the need for repairs was generally recognized. The rocky corners on the western section had been responsible for the sinking of 11 steamers since 1885; and the locks and bridges all needed renovation, no significant work having been done on them since 1911. The cost of essential repairs was put at about £15,000, and that of an improved canal at £700,000.[51]

The passenger-service was finally abandoned in the summer of 1929, when the Ardrishaig–Crinan and Crinan–Oban steamers were replaced by a coach-service between Oban and Ardrishaig; the *Linnet* was wrecked in the Gareloch three years later. A £100,000 renovation scheme was begun in May 1930; the main items were new sea-locks accessible at all states of the tide, a new reservoir to supplement the eight lochs which had so far supplied the canal, and a 15-ft increase in the depth of Ardrishaig harbour. It was hoped that this work would be completed in two years; but it was still in progress in November 1933, and the canal was closed again for a short time the summer after that. The shortage of water remained a problem, and was discussed once again by the British Transport Commission when they took charge of the canal in 1948. The loss on the canal in 1953 was given as £11,999; and the

Commission suggested that responsibility for it should be transferred to the Secretary of State for Scotland. The 1958 inquiry concluded that the canal, though unprofitable, was of greater economic significance than the Caledonian, since it was used by more than twice as many vessels, carried 50,000 tons of goods annually, and formed an important link in the communications between the Clyde and the Hebrides. Shipowners, however, complained that the canal's restricted capacity was detrimental to the economic working of cargo-vessels, and recommended that it be deepened to admit vessels of 10½ ft draught at all times. It was also suggested that the canal might be more extensively used by the Forestry Commission for the shipping of timber from the north-west.[52]

The number of commercial and fishing vessels using the Crinan went down from 963 and 721 in 1951 to 220 and 240 in 1965; but the number of passages made by yachts rose in the same period from 689 to 890. The canal was transferred to the British Waterways Board in 1962; and in 1963, just 150 years after the first passage of Bell's *Comet*, it was used by a hovercraft on its way from Dumbarton to the Thames.[53]

CHAPTER VII

The Caledonian Canal

In 1726 Captain Edward Burt considered the possibility of linking Loch Linnhe and the Moray Firth by a canal through Glen More, but rejected the idea because the channelling of the wind between the mountains would make navigation too 'precarious'. In 1773, however, the Commissioners of Forfeited Estates, who were seeking ways of assisting the fishing industry, asked James Watt to make a survey of the same route; and his report was published in the following year. His plan was for a 10-ft navigable waterway from Fort William to Inverness through Lochs Lochy, Oich, Ness and Dochfour; and he estimated the total cost, with 32 locks, at £164,032. Among the advantages which would result from the construction of such a waterway he noted the establishment of a shorter and safer route from the west coast to the east, the stimulating of the fishing industry, and the provision of cheap corn for the West Highlands. At the same time, he emphasized the difficulties inherent in the scheme, pointing out that vessels might be detained by 'contrary winds' on the lochs, and that time would be lost at each of the canal's numerous entries. The government took no action on Watt's report, but the scheme was not forgotten. John Knox in 1785 reminded the public that a canal like that proposed would relieve distress in the Highlands and enable shipping to avoid the dangerous route through the Pentland Firth, and suggested that it would be of great strategic value in war-time. Another plan was put forward in 1793 by William Fraser, who declared that nature 'not only favoured such an undertaking but had finished more than half of it already'.[1]

There was growing concern at this time about the rate of emigration from the Highlands, and in 1801 the government asked Thomas Telford to investigate the problem. After surveying the area in 1801 and 1802, Telford reported that the principal cause of emigration was the action of landowners in establishing large sheep-farms where the main occupation had hitherto been the keeping of cattle. Recognizing that government interference in this process might be thought inadmissible, he suggested that public works should be carried out in order to 'furnish employment for the industrious and valuable part of people in their own country'; and he declared his belief that the construction of roads, bridges and waterways in the north-west would not only provide immediate employment but also bring about advances in industry, fishery and agriculture. On the scheme for a canal through the Great Glen, he sought the advice of shipowners and others in Leith, Aberdeen, Peterhead, Greenock, Dublin, Liverpool and Bristol; and he found them unanimous about the danger and inconvenience of navigation through the Pentland Firth and the advantages to be expected from the canal. On the navigation of the lochs, he had the advice of Captain Gwynn of the Royal Navy, who reported that Lochs Lochy and Ness were navigable for vessels of any burden and had secure anchorages which could be used when the winds were 'contrary', but that Loch Oich was shallow in some places and would have to be deepened. Telford was confident that Loch Garry and Loch Quoich would provide ample supplies of water, and believed that a canal 20 ft deep could be made in seven years for about £350,000. Besides examining the route from Inverness to Loch Linnhe, he investigated the possibility of linking Loch Eil with Loch Shiel, which had been considered as early as 1766; but in view of the deep cutting that would be required he did not recommend this scheme.[2]

The final incentive for the construction of the canal was the hope of safeguarding shipping from attacks by French privateers. An Act of July 1803 granted £20,000 towards the cost of the project; and its management was entrusted to a board of commissioners including Lord Castlereagh, Lord Binning, Lord Dundas, Sir William Pulteney, Nicholas Vansittart and Charles Grant, who were to meet in London and report annually to Parliament. The canal was to run from a harbour near Inverness 'a little way to the east of the Ferry Pier' to another some distance west of Fort William at the mouth of the Lochy. The Commissioners were to determine the route and negotiate the purchase of land. Disputes about land-prices were to be settled by a jury of fifteen; but no complaint was to hinder the Commissioners from taking possession of the land required. The maximum rates were fixed at 2s per register ton on vessels both laden and unladen and 5s per ton on such goods as were loaded or unloaded in the canal's basins.[3]

The Secretary to the Commissioners was John Rickman; and Telford was appointed principal engineer at a salary of 3 guineas

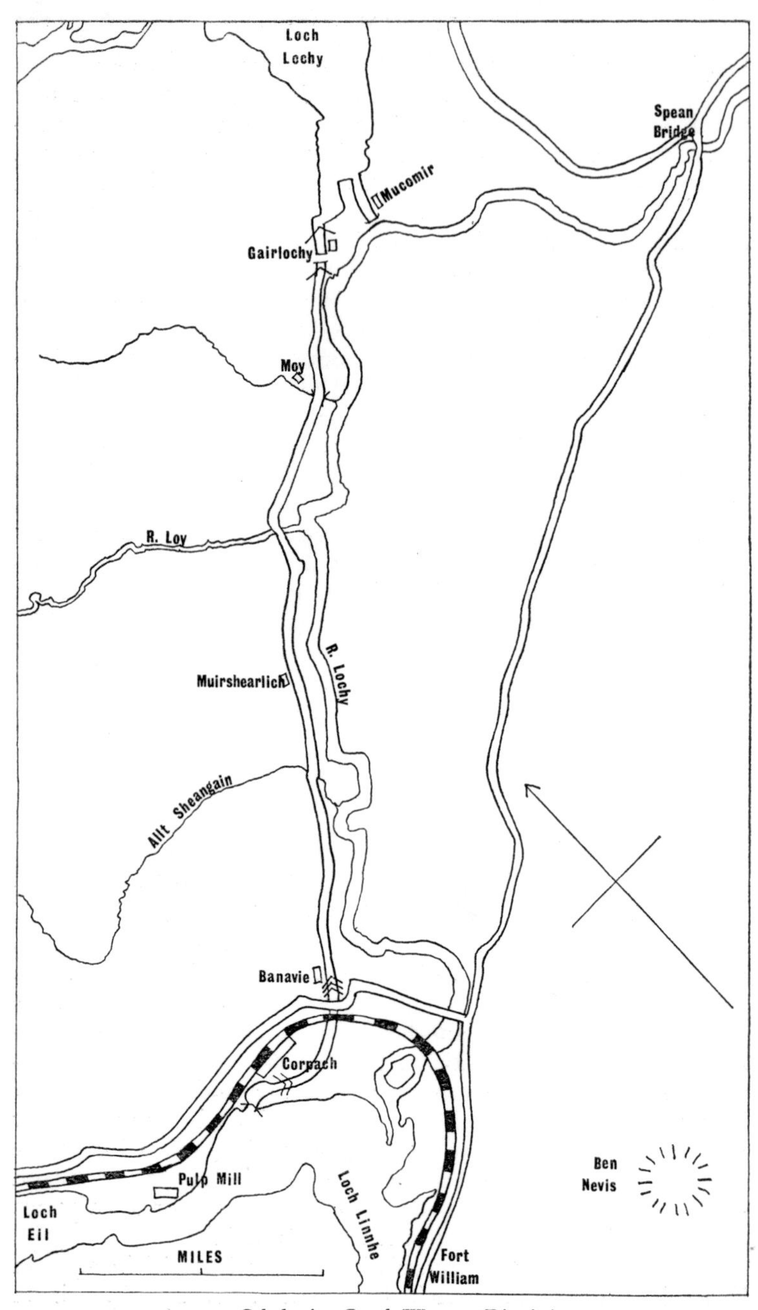

11. Caledonian Canal (Western District)

XV. Crinan Canal: (*above*) the Cairnbaan locks, looking east; (*below*) Islandadd bridge

XVI. Crinan Canal: (*above*) the western end from the air; (*below*) Crinan harbour, looking north

a day plus travelling expenses and instructed to complete his plans as soon as possible. The consulting engineer was William Jessop, for whom Telford had worked on the Ellesmere Canal; and the surveying of the lochs was entrusted to Murdoch Downie. In April 1804 Jessop put the cost of completing the canal at £474,500. This did not include the cost of land, but he believed that prices would be low as the land on the line was not in general of great value. Work began on the terminal basins, each of which was to be 400 yd long and more than 70 yd wide. Turf houses were built for the workmen, and workshops were erected for the blacksmiths and carpenters. Fir and birch were bought locally, and 400 bolls of oatmeal were stored at Corpach. 150 men were employed, including some Highlanders who had worked on canals further south; the wage for labourers was 1s 6d per day, and those who demanded more were 'uniformly rejected'.[4]

An Act passed in June 1804 authorized a grant of £50,000 towards the cost of the canal and made provision for private subscriptions of not less than £50. The southern terminus was moved to a point one mile west of the Lochy, and it was laid down that the northern one should be west of the ferry pier 'at a place called Clacknacarry'. The Commissioners were given authority to deepen and widen all lochs and rivers on the route and to draw water from all lochs, rivers and springs within 5,000 yd of it, including Loch Garry and Loch Arkaig. The maximum rate on goods was given as 2d per ton per mile; but vessels entering a loch were to pay for its entire length. Government ships were to be exempt from dues; and so, for the first three years, were vessels of over 100 tons passing from sea to sea.[5]

On Jessop's advice, Telford appointed two resident engineers or superintendents, Matthew Davidson being made responsible for the works between Clachnaharry and Loch Ness and John Telford (not related to Thomas) for those between Loch Lochy and Corpach; each was to have a house, a horse and a salary of £200 per annum. John Simpson became the contractor for masonry, John Wilson the foreman of the western district and John Cargill the foreman of the eastern district; these men had all worked as masons on the aqueducts at Chirk and Pontcysyllte. It was resolved that the cutting, puddling and embankment should be let in small lots, the Commissioners providing all the materials except picks, shovels and spades. In 1805 it was decided to establish a small brewery at Corpach in order to 'induce the workmen to relinquish the pernicious habit of drinking whisky'; there were also plans for one at Clachnaharry, though this was deemed less urgent, and for the keeping of cows so that the men could be supplied with milk.[6]

The number of labourers had now risen to over 900; but as most of the men were natives of Argyll and Inverness-shire they were sometimes diverted by the claims of the potato-harvest and the herring-fishery. An attempt was made to persuade experienced canal-workers to settle in the district so that their example might 'impart skill and industry' to the other employees. In April 1805 Telford told the Commissioners that the eastern district employed about 500 workers, drawn mainly from the shores of the Moray Firth, and that the western district employed 404, including about 300 'strangers' from Kintyre, Lismore, Appin, Skye, Arisaig and Morar and a number of crofters who had settled on Corpach Moss after being evicted by Cameron of Lochiel from their farms on Loch Arkaig. The poet and novelist James Hogg visited Corpach in 1804, and reported on the progress these workmen had made so far:

> While observing how carelessly the labourers were dabbing with their picks and spades, and how apt they were to look around them at everything which was to be seen, while others were winding slowly out with each a little gravel in a wheel-barrow—while contemplating the exertions of these men, and wishing to anticipate in my mind the important era when they should join Lochiel to the Moray Firth, I could not help viewing it as a hopeless job: my head grew somewhat dizzy, and I felt the same sort of quandary as I used to do formerly when thinking of eternity.

Introducing the Lowland virtues to Glen More was a formidable task; but no doubt the brewery helped.[7]

In order that the locks might admit 32-gun frigates and the largest ships engaged in the Baltic trade, their size was increased in 1805 from 162 ft by 38 ft to 170 ft by 40 ft. This added £8,000 to the estimated cost of the canal; but at the same time the plan for small side-locks to take vessels of 200 tons burden was laid aside. The dimensions of the canal remained unaltered: it was to be 20 ft deep, 50 ft wide at bottom and 110 ft wide at surface. Telford and Jessop proposed that a basin should be made at the Corpach sea-lock, which was to be cut out of rock, and that the locks further inland, each of which was to raise the canal 7 ft 9 in, should be arranged in 'clusters' to save expense. The second and third locks at Corpach were to be united in this way, and so were the next eight at Banavie, which would raise the canal to a level 12 ft above that of Loch Lochy; at the other end there would be a single lock at Clachnaharry and a set of four at Muirtown.[8]

650 tons of 'Memel crown timber' at 2s 4d per cu ft were ordered from a dealer in Aberdeen; and it was resolved that the masonry of the bridges and culverts should be of rubble-stone laid in mortar and that the copings should be of freestone or granite. Cast iron for railways and waggon-wheels was procured from iron works in Aberdeenshire, Denbighshire and Derbyshire; and orders were given for four sloops of 50 tons burden to carry stone, a flat-bottomed boat to carry local timber on the lochs, three Boulton & Watt steam-engines to keep the works dry during the formation of the sea-locks, and another steam-engine to help with the deepening of Lochs Oich and Dochfour. Satisfactory progress was reported in May 1805. Between Muirtown and Kinmylies the canal was almost complete, and work was under way near Torvaine on that part of the canal which was to use the bed of the River Ness. Plans had been made for a channel in the rock south of Fort Augustus for the River Oich; and stone obtained in the making of this channel was to be used for the locks. Digging had begun at Banavie; and at Corpach a timber-framed house had been built for the resident superintendent and the two chief contractors. Progress was limited, however, by the funds available; and the government's annual grant of £50,000 became increasingly inadequate as prices rose and new engineering problems were encountered.[9]

The number of labourers at work in 1805 and 1806 varied between 641 and 1,063, being reduced sometimes by shortage of money and sometimes by the attractions of the herring-fishery, the potato-harvest and the cutting of peat. Applicants rejected when funds were short were assured that they could easily get employment on the new roads and would find this less arduous than canal work. Because of the risk to the masts of vessels, Telford and Jessop decided against building drawbridges like those on the Forth & Clyde; instead, they proposed to have cast-iron swing-bridges like those at the West India Docks. Muirtown Basin, which had been replanned to measure 800 yd by 140 yd, was almost completed in 1807. For financial reasons, work was concentrated at the eastern and western ends of the canal, the rock-cutting near Fort Augustus being discontinued. As a consequence the Commissioners received a petition from more than 300 tacksmen, tenants, cottars and labourers of the middle district, who asked that work be recommenced there, since many of them would otherwise be obliged by lack of employment to 'seek for subsistence at a distance and desert their native country'. No action was taken on this petition; but the Highlanders who were employed received regular four-weekly payments, and it was claimed that this procedure resulted in an 'increased assiduity' which kept labour costs down.[10]

The 1808 report was optimistic about supplies of stone: rubble-stone had been discovered near Banavie, limestone was available in Lismore, and the small quantities of freestone needed were being shipped from the Cumbraes. A railway was established at Corpach, and a stable was built to accommodate the horses used on it by the masonry contractors. It was decided that the gates for the second and third locks at Corpach should be of American pitch-pine sheathed in cast iron. The cast iron for the bridges and lock-gates south of Loch Lochy came from Hazeldine's foundry at Plas Kynaston in Denbighshire, and was brought by canal to Chester and then by sea to Corpach; that for the eastern and middle districts came from the Butterley foundry, of which Jessop was part-owner, and was brought by canal and river to Gainsborough and then by sea to Inverness. The general supplanting of Baltic by North American timber caused by the Napoleonic blockade was to affect the canal not only by increasing costs and introducing inferior materials but also by reducing that trade to Northern Europe from Liverpool, Belfast and Glasgow which had been regarded as a major source of future revenue.[11]

The 'extraordinary augmentation of the price of timber' in 1808 and 1809 made it necessary to cut labour costs and postpone work on the middle district. Quarries were opened above Corpach moss to supply rubble-stone for the chain of locks being built at Banavie; and stone for the eastern district was quarried at Clachnaharry and at Redcastle on the Black Isle. Timber prices delayed the building of a coffer dam for the Clachnaharry sea-lock; and because the shore at the northern entry consisted of soft mud a new method of construction was decided on. Matthew Davidson, who was in charge of operations at Clachnaharry, resolved in 1809 to bring a large quantity of 'rubbish' from the nearby quarry to form a solid mass in which the lock-pit could be dug; and in order to transport material to this embankment and to the locks at Clachnaharry and Muirtown, a short railway was built with iron partly from the Butterley foundry and partly from the Leys Ironworks of Aberdeen.[12]

The work between Corpach and Loch Lochy was complicated by the need to control the water from mountain streams which after heavy rain could become torrents. Culverts at Loy, Muirshearlich, Sheangain, Upper Banavie and Lower Banavie were made in 1809, and were expected to save the cost and inconvenience of bridges by permitting carts and cattle to pass under the canal. Three of the eight locks known as 'Neptune's Staircase' were finished by June 1809; and it was hoped that the whole chain would be ready the following year. Because of labour difficulties, however, only six of the locks were completed in 1810; and it was not until the end of 1811 that this 'great mass of masonry extending 500 yd in length' was considered perfect. By this time, the labour shortage in the western district had been alleviated by a 'temporary embarrassment among the manufacturers in Glasgow and its vicinity', and about 1,200 men were at work on the canal; the Commissioners were able to congratulate themselves on having afforded 'seasonable resource' to the unemployed.[13]

The excavation of the lock-pit at Clachnaharry in 1810–11 was a formidable task. The influx of sea-water proving too much for the 'ordinary means of clearance', hand-pumps were replaced by a large chain-pump worked by six horses; and when the strain began to tell on the horses one of the canal's 6 hp steam-engines was brought into 'powerful but expensive' operation. By this means, the pit had been sunk by May 1811 to within 4 ft of the required depth; and it was then resolved that the gates for the sea-lock should be make of oak from North Wales. The northernmost stretch of the canal was enclosed by two mounds extending for 400 yd from the sea-lock to the high-water-mark, where the second or Clachnaharry lock was situated. Above this lock was the Muirtown Basin, intended primarily for the Inverness trade; and the canal was now completed, except for the embankment over the Kinmylies culvert, to a point five miles from the northern entry.[14]

Operations in the western district had by this date reached the foot of Loch Lochy, where a new channel for the River Lochy was being made so that part of the old river-bed could be incorporated in the canal. Because of the 'slow progress and great expense' involved in excavating the Corpach basin in 'solid rock', it was decided that it should be only 250 yd long and 100 yd wide. A 20 hp steam-engine was sent by canal and sea from the Soho works of Boulton & Watt, and was erected at Corpach under the supervision of one of the firm's fitters; and it was hoped that the sea-lock, whose gates were to be of Welsh oak, would be finished by July 1812. With less than half the canal finished, however, expenditure had mounted to nearly £343,000; and in February Telford was cautioned against engaging too many workmen.[15]

Work began in the middle district in September 1811; 350 men were employed, half of them being accommodated in huts specially erected by the Commissioners. Progress was slow, however, because there were doubts about the best way of connecting the canal with Loch Ness and uncertainties about its course across the government's land at Fort Augustus, where five locks had to be built. Some work had been done earlier on a new channel for the River Oich; but this was still unfinished in June 1813. By that time, however, considerable progress had been made in the eastern and western districts. Eighteen locks were completed, the single ones being 170 ft and the staircase ones 180 ft long and both types 40 ft wide. At the eastern end, work had concluded on the first and second locks at Clachnaharry, the 20-acre Muirtown Basin and the staircase of four locks above it, and the two sections of canal bank in the River Ness; and it was hoped that the regulating-lock near the east end of Loch Dochfour would be ready in 1814. At the western end, there was 18 ft of water in the sea-lock and 10 ft in the Corpach basin; and the canal from there to Lower Banavie was finished and had been filled to a depth of 5 or 6 ft. At Gairlochy, where the regulating-lock was finished except for the freestone coping, a house with stables had been built for the workmen and horses.[16]

In November 1812 William Grant, who had been a labourer on the canal, brought accusations of fraud against Alexander Easton, the former Forth & Clyde mason who had succeeded John Telford as engineer of the western district in 1807, and against Easton's agent Angus MacDonald. Grant alleged that Easton had collaborated with the master workmen in mis-measuring the volume of earth dug, and that MacDonald had 'bestowed 200 yd on me and upwards for my treating him with a few bottles of porter'. These charges, however, were dismissed by a court of inquiry. Easton was given permission to dismiss all workmen who tried to 'excite others to discontent'; and the workmen were told they should in future present their complaints to the local magistrates. In May 1814, Easton having complained that his income was 'insufficient for maintaining a due respectability in his situation', his salary was raised to £250 per annum. Davidson's salary was raised at the same time to £300; and Telford's allowance for travelling expenses was increased from 1s 6d to 2s per mile.[17]

In a report of October 1813 Telford suggested that the plan for a towpath along Loch Oich should be abandoned, since the use of steamboats like those which had begun to ply between Glasgow and Greenock would render horses unnecessary for towing purposes. To account for the high rate of expenditure on the canal, Telford pointed out how costs had risen since the work began.

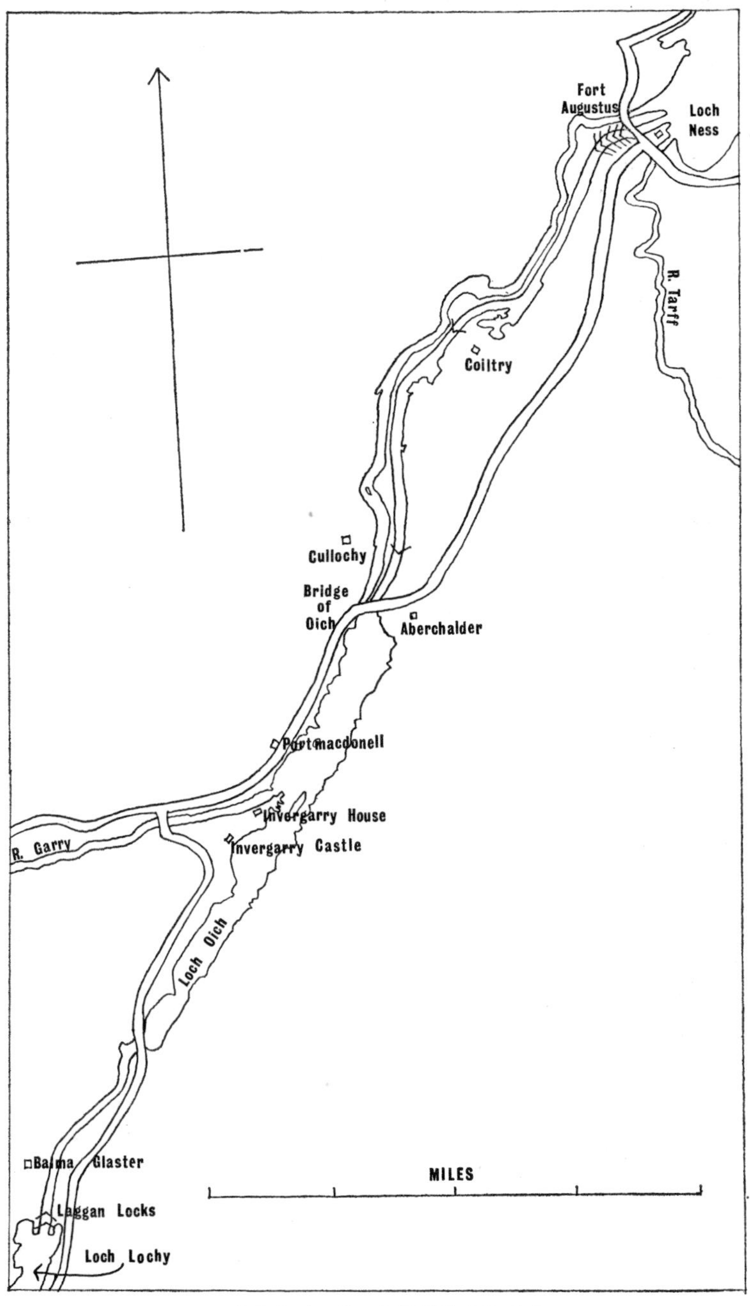

12. Caledonian Canal (Middle District)

Labourers' wages, he said, had gone up from 1s 6d to 2s 6d per day, blacksmiths' wages from 2s 6d or 3s to 3s 6d or 4s per day, and masons' wages from 16s to 21s per week; and piece-work rates had gone up from 3d to 4½d per cu yd. The price of oatmeal had risen from 20s or 21s to 38s per boll, that of a horse in the western district from £25 or £30 to 40 or 50 guineas, that of rope from 75s to 100s per cwt, and that of Baltic timber from 2s 6d to 7s per cu ft. Telford put the cost of completing the work at £234,734, and believed that if the annual grant was increased the canal 'might be opened for all purposes of commerce at the end of 1817'.[18]

Sir Walter Scott, who visited Loch Linnhe in August 1814, was pessimistic about the canal's future:

> Had the canal been of more moderate depth, and the burdens imposed upon passing vessels less expensive, there can be no doubt that the coasters, sloops and barks would have carried on a great trade by means of it. But the expense and plague of locks etc may prevent these humble vessels from taking this abridged voyage, while ships above 20 or 30 tons will hesitate to engage themselves in the intricacies of a long lake-navigation, exposed without room for manœuvring to all the sudden squalls of the mountainous country.

Progress with the regulating-lock at Dochgarroch had been delayed by the need to bring materials by land from Clachnaharry and by the influx of water through the bank separating the canal from the River Ness; but the lock was completed in 1814. Further south, an appropriate site for a staircase of five locks was sought in the coarse gravel at Fort Augustus; and plans were made for raising the level of Loch Lochy by 12 ft so that it could be linked with the middle and western sections of the canal.[19]

There was at this period a serious dispute between the Commissioners and Colonel Alexander Ranaldson MacDonnell of Clanronald and Glengarry, who lived in a mansion-house built near the ruins of Invergarry Castle on Loch Oich and owned land extending for 10 miles along the Great Glen. Glengarry had originally opposed the canal as a threat to his privacy, but had accepted an assurance from Telford that its route would follow the southern shore of Loch Oich and be separated from the loch by an embankment. The value of the ground to be acquired from him was settled by a jury at £10,000, and of this sum £2,000 was paid in 1814 and the remainder in 1816; but Glengarry, who was suffering like other Highland landowners from the post-war decline in agricultural prices, maintained a vigorous opposition to the Commissioners' work on Loch Oich, claiming that they would ruin the fisheries there. Early in the morning of 3 September 1816, we are told:

> Glengarry came to the east end of Loch Oich accompanied by about thirty persons variously armed as if for deer-hunting, who drove away the workmen and having seized a boat belonging to the Commissioners sent it to Loch Garry.

Evidence of this 'outrage' was submitted to the Sheriff Depute and forwarded by him to the Lord Advocate; but the Lord Advocate declined to prosecute Glengarry as a public offender, and the Commissioners, having recovered their boat, decided to take no further action.[20]

The work of dredging Loch Oich continued with the help of two of the earliest steam bucket-dredgers; these had been built at the Butterley foundry to the plans of Bryan Donkin, who also advised Telford on the possibility of using steam-tugs on the lochs. The cost of completing the canal was estimated in 1816 at £170,000; and later in the year the Commissioners received a grant of £75,000 instead of the usual £50,000 so that they could buy cast iron for the lock-gates. John Simpson, who had been the canal's principal contractor, died in June 1816, when the work on the middle district (later to be much criticized) was still in its early stages.[21]

Problems arose in the excavation of the lock-pits at Fort Augustus: the River Oich having been diverted to the north side of a small island so that three of the five locks could be built in the old river-channel, water kept filtering through the gravel of the island so that 'even a steam-engine pit' could not be sunk without great difficulty. The sill of the bottom lock had to be fixed 20 ft below the lowest water-level of Loch Ness, and three steam-engines, including one of 36 hp, were employed to achieve this. Robert Southey inspected the works at Fort Augustus on 16 September 1819, and wrote in his journal:

> Such an extent of masonry, upon such a scale, I had never before beheld, each of these locks being 180 ft in length. It was a most impressive rememberable scene. Men, horses and machines at work, digging, walling and puddling going on, men wheeling barrows, horses drawing stones along the railways. The great steam-engine was at rest, having done its work; but the dredging-machine was in action, revolving round and round, and bringing up at every turn matter which had never before been brought to the air and light. Iron for a pair of lock-gates was lying on the ground, having just arrived from Derbyshire.

The five locks were not completed until 1820.[22]

The government grant for 1817 was only £25,000; but this reduction had no effect on the progress of the work. In the following year the Commissioners suggested that the disused military establishments at Fort Augustus and Fort William might be transferred to them for use as warehouses; but this proposal was rejected in 1819 by the Duke of Wellington, and the forts remained under the control of the Ordnance Board. A more serious problem was created by the refusal of the military authorities to permit any interference with the glacis of the fort at Fort Augustus; but this refusal was withdrawn in May 1819, and the passage through the glacis was completed a year later.[23]

Having found a total separation of the canal from Loch Oich financially impracticable, Telford engaged in further negotiations with Glengarry in 1818 and 1819. His proposal that a channel should be made along the southern side of the loch was rejected, however; and the Commissioners, deciding that agreement with Glengarry was unattainable, gave orders that the navigation should follow the direct route through the centre. Two years later, when Glengarry was holding up the opening of the canal by preventing the masons from working on the regulating-lock at Cullochy, the Commissioners pleaded with him to obviate recourse to a jury by withdrawing his objections. They were unwilling to submit landowners' claims to local juries because these juries were biased against the government interest and tended to award 'extravagant amounts of compensation'. In this case the personal appeal was successful; but the hope of altogether avoiding litigation with Glengarry proved illusory.[24]

The main problem on the canal's summit level in the Laggan district between Loch Oich and Loch Lochy was the depth of the 2-mile cutting required. According to Southey, who watched the excavation being carried out in 1819, the earth was removed 'by horses walking along the bench of the canal, and drawing the laden cartlets up one inclined plane while the emptied ones were let down by another'. As soon as a part of the cut was deep enough to be flooded, the Donkin dredgers were brought in to increase the depth. A staircase pair had to be built at the west end of the Laggan cut; and this was completed in 1822. In May of that year Glengarry 'unexpectedly' demanded a bridge over the Laggan cut near Balma Glaster to serve the village he had established round a new episcopalian chapel; but as he had already been awarded four bridges the Commissioners denied any obligation to build a fifth. They were equally unsympathetic when Cameron of Lochiel renewed a 'long dormant' claim for damage resulting from the separation of part of his land from the nearest farmsteadings.[25]

With the collapse of the Baltic timber-trade, the end of the Napoleonic Wars and the development of steam navigation, public enthusiasm for the canal declined; and in May 1818 Telford tried to explain why the construction was taking so much longer and costing so much more than he had predicted. He placed most emphasis on the 30 to 50 per cent rise in the price of food, materials and labour and the series of bad harvests which had made wage-reductions impossible; but he also pointed out that the cost of land and quarries had exceeded his expectations, that some parts of the canal had had to be lined with clay to prevent injury to the adjacent lands, that the lock-gates had had to be made of cast iron instead of oak, that more rock-cutting had been required in the western district than had been foreseen, and that the assembling of dredgers in 'that remote country' had proved surprisingly costly.[26]

A grant of £50,000 was made in 1818; but in the following year there were suggestions that the canal might have to be abandoned for lack of funds. 'How galling', wrote William Young in a private letter of 13 March, 'that the Caledonian Canal works are to be stopped. Scotland at large should join in a petition to the House craving aid.' In the Commons that month Lord Carhampton, having criticized the canal's name on the ground that 'the race of the Caledonians had been extinct for 600 years', declared that the work had been 'originally devised for the purpose of employing the labourers of Scotland' but that 'all that benefit had been reaped by Ireland'. W. Smith, on the other hand, maintained that the canal would be 'productive of the greatest benefits', that 'the spirit of labour' which had been introduced into the Highlands was 'worth more than all the money that had been voted', and that among all those employed 'there were not two in a hundred who were not Scotchmen'. The canal had now cost £700,000, but it was estimated that it could be completed for a further £80,000; and the House therefore agreed to continue the grant to the Commissioners for a further year.[27]

The Commons debate initiated a vigorous controversy. In their 1819 report the Commissioners declared that they had 'given no directions nor exerted any influence for the accomplishment of an invidious distinction between the natives of a United Kingdom', but that the proportion of non-Highland labour on the canal was very small. Telford gave the proportion of 'strangers' among the workers in the eastern district as 1 in 70 in the years 1804 to 1808,

1 in 186 in 1809, 1 in 80 in 1811 and 1812, 1 in 46 in 1813, 1 in 64 in 1815 and 1817, 1 in 70 in 1818, and 1 in 48 in 1819. An article in *Blackwood's Edinburgh Magazine* defended Telford against his critics, describing the canal as 'one of the most magnificent and splendid of our national structures' and declaring that it had been undertaken 'for the general good and commercial enterprise of the United Kingdom' and had already brought considerable benefits to the north-west:

> Prior to the period of the commencement of the Caledonian Canal, the inhabitants of the Highlands and Western Isles of Scotland were in a state which may be described as maintaining a degree of apathy even beyond what has been too often ascribed to them. But let anyone now traverse these mountainous and insulated districts to the northward of that chain of lakes forming the track of this canal, and he will be astonished at the change which has been produced even within the last ten or twelve years upon the intelligence and manners of the inhabitants.

Even *Blackwood's*, however, had to admit that 'there might be no immediate use for a canal of a large capacity in this situation'.[28]

Fierce criticism of the canal continued in Parliament and elsewhere; but in 1820 the Commissioners received a grant of £60,000. Loch Ness and the eastern district had been opened for navigation in May 1818; and coasting vessels importing tar, oatmeal and coal and exporting wool, staves and timber had made 150 voyages in the first summer. By September 1819 the practicability of the voyage from Inverness to Fort Augustus had been proved 'even by some square-rigged vessels of 140 tons'; and in November 1820 Henry Bell established a steamboat service on this part of the navigation. The *Stirling Castle*, which was 68 ft long and 23 ft wide and had an 18 hp engine, left Muirtown at 8 am and reached Fort Augustus at 2 pm. A diligence carrying four inside and three outside passengers proceeded from Fort Augustus to Fort William; and from there another steamboat took passengers via the Crinan Canal to Greenock and Glasgow. The locks at Corpach had been in operation since 1819; and Telford reported that vessels of 100 tons burden had loaded and unloaded cargoes in the basin.[29]

The Commissioners received no grants in 1821; but grants of £25,000 were made in 1822, 1823 and 1824. In 1822, because of the attacks in Parliament, it was decided that the canal should be opened at once despite the inadequate depth of some sections. The formal opening took place in October 1822, Glengarry having 'collected his men and armed them several days previous to being invited to join the party'. On the 23rd a group of local landowners and others set out in two boats on a two-day voyage from Inverness to Fort William, which was expected to shame and silence 'the doubters, the grumblers, the prophets and the sneerers'. The *Inverness Courier* reported:

> Amid the hearty cheers of a crowd of spectators and a salute from all the guns that could be mustered, the voyagers departed from the Muirtown locks at 11 o'clock, with fine weather and in high spirits. In their progress through this beautiful navigation they were joined from time to time by the proprietors on both sides of the lakes; and, as the neighbouring hamlets poured forth their inhabitants, at every inlet and promontory, tributary groups from the glens and braes were stationed to behold the welcome pageant, and add their lively cheers to the thunder of the guns and the music of the Inverness-shire militia band, which accompanied the expedition. . . . The reverberation of the firing, repeated and prolonged by a thousand echoes from the surrounding hills, glens and rocks, the martial music, the shouts of the Highlanders and the answering cheers of the party on board produced an effect which will not soon be forgotten by those present.

After a night at Fort Augustus, the company entered Loch Oich early on the 24th:

> On approaching the mansion of Glengarry, the band struck up 'My name it is Donald Macdonald', and a salute was fired in honour of the chief, which was returned from the old castle, the now tenantless residence of Glengarry's ancestors. The ladies of the family stood in front of the modern mansion waving their handkerchiefs.

Glengarry having joined the party, the two boats, now accompanied by *Comet II*, proceeded through the Laggan cut and down Neptune's Staircase to Fort William.

> The termination of the voyage was marked by a grand salute from the fort, whilst the inhabitants demonstrated their joy by kindling a large bonfire. A plentiful supply of whisky, given by the gentlemen of Fort William, did not in the least tend to damp the ardour of the populace.

At a dinner given that evening by Charles Grant, one of the Commissioners, the canal was praised as 'one of the most stupendous undertakings of that nature which Europe had seen', and it was claimed that the people of the Highlands had 'received a new set of ideas' by seeing the 'skill, ingenuity and expertness' with which

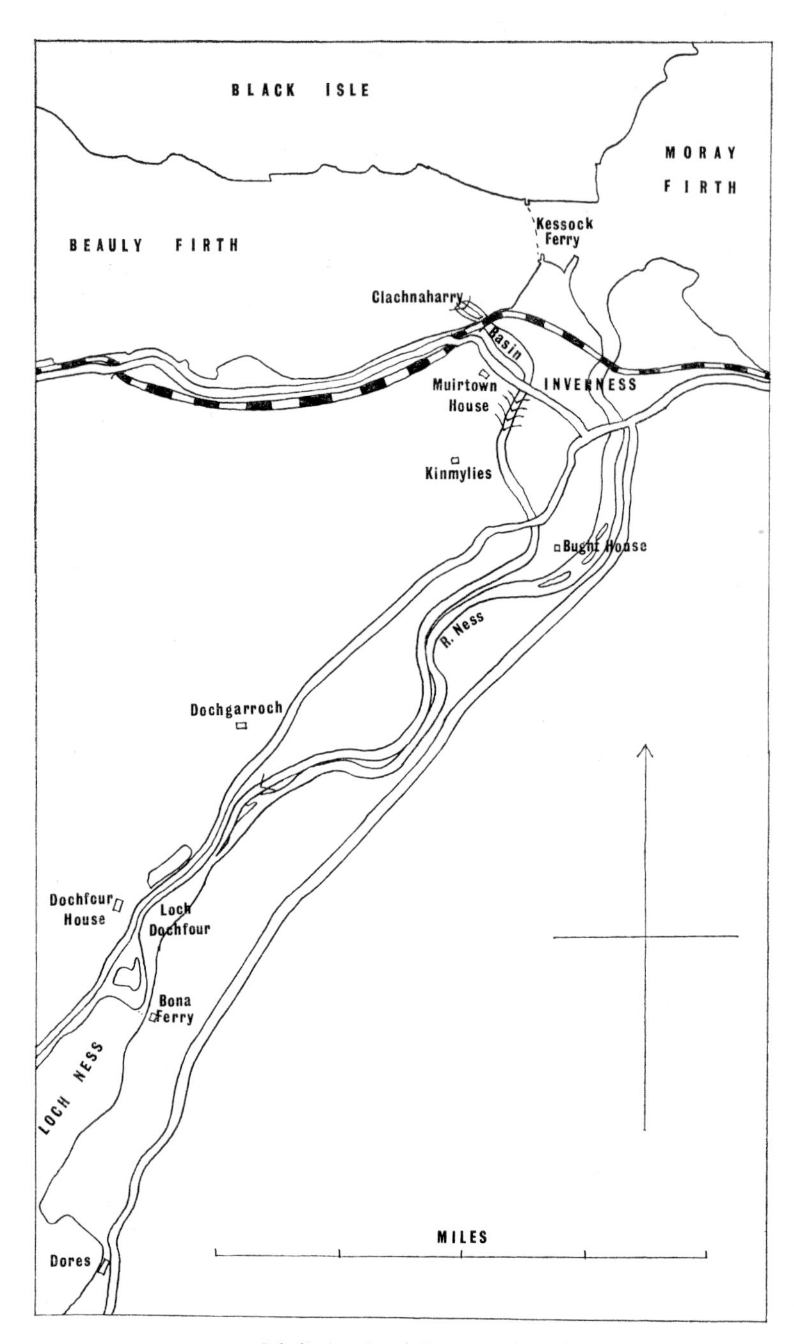

13. Caledonian Canal (Eastern District)

the work had been done. Glengarry, however, in proposing Grant's health, expressed the hope that 'the better part of the ancient spirit of the Highlanders would not be destroyed by the improvements of modern days'. In a private letter, Easton reported to Telford that

> the evening passed over with much mirth and pleasantry, excepting a slight controversy betwixt Glengarry and one of the gentlemen from the east as to chieftainship, which through Mr Grant's complacency was smoothed over.

On the 25th, the company set out on the return journey to Inverness; many of the landowners, however, left the boats en route, 'each returning to his hereditary mansion and subject valley'.[30]

Despite his presence at the formal opening of the canal, Glengarry still resented the fact that it traversed his property. In a letter of 30 April 1823 he complained that the 'privacy and retirement' of his mansion were destroyed now that 'passage-boats and smoking steam-vessels' were passing daily through Loch Oich; and he demanded compensation 'for the loss of amenity occasioned by such a fine sheet of water in my pleasure-grounds being laid open to the public'. One of the vessels complained of was the *Stirling Castle*, which had begun in 1822 to operate between Muirtown and Banavie.[31]

Tonnage dues in 1823 were ¼d per ton per mile, steamboats carrying only passengers and light parcels being charged 10s each for the entire sea-to-sea passage. Lock-keepers were required to give priority to passenger-boats, and navigation was prohibited during the night. Horses were available for tracking vessels at a speed of 2 mph and a charge of 5s per day. Easton resigned in 1823, having 'obtained a situation in Ireland'; and it was resolved in view of 'the approaching completion of the work' that no successor should be appointed. James Davidson, who had succeeded his father as superintendent of the eastern district in 1819, thus became resident engineer for the whole canal at a salary of £300 per annum.[32]

Between 1 May 1823 and 1 May 1824 the canal was used by 844 vessels, of which 278 passed from sea to sea; and by the summer of 1824 one of the lock-houses at Banavie had been converted into an inn for the benefit of the steamboat passengers. Few sailing-boats had so far needed to be towed on their way through the canal: the average time for the passage had been three to four days, and there had been no complaints about 'detention' by unfavourable winds. The canal's revenue for 1823–4 was £2,159; and it was reported that 466 complete and 675 partial passages had been made. Three steamboats were now in operation between Glasgow and Inverness, doing the return journey in six days; and two sailing-boats had begun to ply between Liverpool and the Moray Firth. By November 1825 the canal had been deepened to 15 ft throughout.[33]

Glengarry having petitioned the Sheriff of Inverness for compensation on the ground that Loch Oich had been opened to navigation, the Commissioners feared that similar claims might be made by 'all persons who resided within sight of vessels passing along the canal or lakes'. In 1825, therefore, an Act was passed requiring all claims for damage resulting from the creation of the canal to be submitted by 1 February 1826. Tonnage dues were increased on 1 July 1825 from ¼d to ½d per ton per mile, partly in the hope of increasing revenue and partly to answer complaints of unfair competition which had been made by the Forth & Clyde. Revenue for 1826–7, however, was only £2,445; and it was reported that many shipowners were using the Pentland Firth rather than pay 2s 7d per ton for the passage of the canal. In 1828, therefore, the original tonnage rate of ¼d per ton per mile was re-established.[34]

The 1824 grant was exhausted by 1826; and the only money then available for current expenses was the accumulated revenue, which in February 1826 stood at £3,345. The Gairlochy regulating-lock was said to be in a 'dangerous state', and it was estimated that it would cost £1,577 to repair. Telford attributed the premature decay of this lock to the fact that it had been made with inferior stone, no good quality stone having been available on Loch Lochy and the canal from Corpach not having been open for the transport of stone from further afield. On 22 March part of the side-wall of the lock collapsed; and the Commissioners gave orders for its immediate reconstruction. The work cost £3,314, and was completed on 8 June.[35]

By April 1827, the Commissioners had overdrawn £1,686 from the Bank of Scotland; and in a letter to the Treasury they drew attention to the 'embarrassing state' of the canal's affairs. Lord Colchester was sent to investigate; and in May 1828 he reported that the canal was being administered economically and efficiently, that minor defects were promptly remedied by the lock-keepers, and that the labourers who maintained the banks were chosen for their 'attentive habits and active dexterity'. Revenue for 1827–8, which included four months on the reduced tonnage dues, was still only £2,500. The Commissioners, however, described the canal as a great 'future resource', maintaining that its value would

XVII. Caledonian Canal: (*above*) Corpach basin, with the pulp mill beyond; (*below*) looking south from Banavie locks

XVIII. Caledonian Canal: the locks at Banavie, with Loch Lochy beyond

be proved when the 'superior quality' of Baltic timber caused trade to 'revert to its natural channel'. They suggested, therefore, that a small grant should be made in order to 'prevent the dilapidation of the magnificent apparatus of this national work'.[36]

The canal's 'magnificence' as a feat of engineering was the theme of three blank-verse 'inscriptions' published by Southey in 1829. The longest of these, which was designed for a site in the centre of the glen, celebrated the 28 locks, the culverts by which rivers and animals passed under the waterway, the man-made channels in lochs and river-beds, and the 'massive outwork' which had been created for the sea-lock at Clachnaharry:

> Thou who hast reach'd this level where the glede,
> Wheeling between the mountains in mid air,
> Eastward or westward as his gyre inclines
> Descries the German or the Atlantic Sea,
> Pause here; and, as thou seest the ship pursue
> Her easy way serene, call thou to mind
> By what exertions of victorious art
> The way was open'd. Fourteen times upheaved,
> The vessel hath ascended, since she changed
> The salt sea water for the highland lymph;
> As oft in imperceptible descent
> Must, step by step, be lower'd, before she woo
> The ocean breeze again. Thou hast beheld
> What basins, most capacious of their kind,
> Enclose her, while the obedient element
> Lifts or depones its burthen. Thou hast seen
> The torrent hurrying from its native hills
> Pass underneath the broad canal inhumed,
> Then issue harmless thence; the rivulet
> Admitted by its intake peaceably,
> Forthwith by gentle overfall discharged:
> And haply too thou hast observed the herds
> Frequent their vaulted path, unconscious they
> That the wide waters on the long low arch
> Above them, lie sustained. What other works
> Science, audacious in emprize, hath wrought,
> Meet not the eye, but well may fill the mind.
> Not from the bowels of the land alone,
> From lake and stream hath their diluvial wreck
> Been scoop'd to form this navigable way;
> Huge rivers were controll'd, or from their course
> Shoulder'd aside; and at the eastern mouth,
> Where the salt ooze denied a resting place,
> There were the deep foundations laid, by weight

> On weight immersed, and pile on pile down-driven,
> Till steadfast as the everlasting rocks
> The massive outwork stands. Contemplate now
> What days and nights of thought, what years of toil,
> What inexhaustive springs of public wealth
> The vast design required; the immediate good,
> The future benefit progressive still;
> And thou wilt pay thy tribute of due praise
> To those whose counsels, whose decrees, whose care,
> For after ages formed the generous work.

In the poem which was to be inscribed beside Neptune's Staircase at Banavie, Southey eulogized the engineer who had designed not only this canal but also the Pontcysyllte aqueduct and the Menai Bridge; and in the poem to be inscribed at Clachnaharry he recalled how the canal had been begun 'in a time of arduous war' and completed

> When national burdens bearing on the state
> Were felt with heaviest pressure,

and prophesied that it would be ranked 'in days to come' with the greatest public monuments of Egypt, Greece and Rome.[37]

Because of serious faults at Banavie and defects in the locks elsewhere, navigation had to be stopped for a fortnight in April 1829. Revenue fell in 1828–9 to £2,194; but this was attributed to the general depression of trade. James Davidson, having been advised for the sake of his health to seek a warmer climate, remitted his responsibilities to George May, the toll-collector at Clachnaharry; and when it became clear that Davidson would be unable to return, May became resident engineer. In July 1830, Samuel Smith succeeded John Rickman as Secretary to the Commissioners. According to May, the canal was now in better condition than at any time since the opening, and there was no reason to expect that any serious repairs would be required in the immediate future. There were demands, however, for steam-tugs on the lochs; and in July 1830 it was agreed that a Mr Stevenson should be allowed free of dues to operate a small steamboat with a 6 or 7 hp engine using it partly for his own purposes and partly to tow sailing-vessels.[38]

Outstanding claims against the canal amounted in 1831 to £39,146; but nearly half of this was for damage done to the salmon-fishery, which the Commissioners believed to be grossly exaggerated. Glengarry's claim had been taken over by his son Aeneas; and it was agreed that 'some compensation' was due in this case. In spite, however, of a government grant of £4,886 in

1829, the Commissioners still owed £5,000 to the Bank of Scotland; and the canal's annual expenditure was about £1,600 less than its annual income. There was little hope, therefore, of an immediate settlement; and in fact negotiations were not completed until 1846. The Commissioners hoped that the canal's depth could be increased to 20 ft, and that a reduction in the duty on Baltic timber would enable northern Europe to regain the dominant position in the timber trade which it had lost to North America during the war.[39]

In June 1834 the Commissioners received a memorial in which eighty merchants and shipowners of the Moray Firth asked that proper arrangements should be made for the towing of sailing-vessels on the canal so as to remove the risk of 'detention' which had hitherto discouraged use of the navigation. Since a tug with a 40 hp engine would have cost 'at least £2,000', the Commissioners decided to enlist the aid of 'some respectable individual' who would 'undertake the speculation on his own risk'; but when they applied to the shipowner responsible for the tug-service at Aberdeen harbour they met with a refusal. Reporting in September 1835, James Loch declared government aid in the introduction of tug-boats to be essential if the canal was to transcend its 'present limited condition'. He was critical, too, of the accommodation provided for passengers, declaring the inn at Banavie (which had been described in the previous year as 'excellent') to be too small and too far from the sea-lock.[40]

Romantic enthusiasm for the canal was still alive in the 1830s, as a guide-book entry demonstrates:

> A spectacle more gratifying to every patriotic feeling can hardly occur, than to see stately ships which the day before had been surmounting the billows of the ocean, and loaded with the produce of foreign climates, sailing on the placid lochs of Caledonia, under the brow of her lofty mountains, or gliding along the lake, whose banks are covered with corn and cattle, while the seamen are cheered with the rustic lay of the shepherds, or the sounds of rural industry within her peaceful valleys.

In the same decade, however, the canal's defects were being brought to notice. In November 1834, after heavy rain which had raised the level of Loch Ness by 2 ft in a single night, the Mucomir outlet of Loch Lochy was blocked and the water-level in the loch rose to 3 ft above the top of the Gairlochy lock-gates; only the united efforts of the lock-keepers and the crews of the Glasgow and Liverpool steamboats prevented the flood from breaking the canal-banks. In June 1836 the wing-walls and gate-recesses at the foot of the Fort Augustus locks were reported to be in a very defective state; and in January 1837 J. A. Borron of Strontian wrote to Lord Lovat drawing attention, among other things, to the inconvenience caused by the inadequacy of the lock at Gairlochy.[41]

In November 1837 George May replied to Borron's criticisms by giving a detailed history and survey of the canal. After recalling the motives and expectations of the original promoters and describing the navigational difficulties which had been encountered since the opening, he pointed out what he regarded as the central flaw in the promoters' reasoning:

> The circumstances connected with the physical formation of the great valley, namely, its consisting for by far the greater part of its length of inland lakes and arms of the sea, which were considered to hold out peculiar inducements, and to offer singular facilities for the construction of a navigation on a large scale, are precisely those which have been found in practice to occasion the most material obstacles to its success and general usefulness.

Seeking an explanation for the way in which 'the difficulty of navigating large vessels through narrow lakes and estuaries' had been overlooked, he concluded that 'the attractive splendour of the design, and the imposing structure of the valley' must have 'outweighed every prudential consideration'. While he doubted the wisdom of the original plan, however, May contended that 'the fortuitous progress of events' had produced in steam navigation an unforeseeable means of justifying the canal's existence; and after pointing out that he had reported on the need for steam-tugs some years earlier, he gave his views on the type and number of vessels which would be appropriate for this purpose. In his survey of the canal, he expressed himself satisfied with the state of the Clachnaharry entry, the Muirtown basin and locks, the Dochgarroch lock and the sea-lock at Corpach; but he drew attention to many defects in the middle section, where extensive use had been made of inferior stone and timber, and spoke still more harshly of the great staircase at Banavie:

> The masonry throughout the whole structure, so far as I have had the means of becoming acquainted with its internal condition, and judged with reference to the purposes for which it was intended, I cannot characterize by any other term than that of execrable; and indeed I have reason to believe that the contractor for these locks, while engaged in the actual execution of the

work, was fully impressed with the conviction (which was shared by many others at the time) that the navigation was a thing which was never to take effect, and that his locks would consequently never require to come into actual operation. That they have in some measure been enabled to do so successfully, is entirely owing to the fortunate nature of the foundations; for had such masonry stood on the ground on which the Fort Augustus locks are built, it is my opinion that the Banavie locks would at this moment have been an entire heap of ruins.

The fact that Telford had accepted such workmanship May attributed to the trouble which the contractor had taken 'to conceal the true nature of his proceedings'.[42]

In December 1837 the north-west recess wall of the bottom lock at Fort Augustus gave way; and May decided that all the masonry beside the lower gates would have to be taken down and rebuilt. One of the gates having been forced shut by the falling stones, there was room for only the narrower boats to pass through the lock. On learning of this, the Treasury asked James Walker to investigate the canal's condition and prospects and determine what repairs were needed. In his report Walker listed as essential to safety the renewal of the Fort Augustus locks, the provision of a new weir for releasing flood-water from Loch Oich, and the construction of an additional lock at Gairlochy; and he estimated the total cost of this work at £24,827. He pointed out, however, that this would still leave the canal in 'a very unfinished state'; and he went on to list the further repairs and improvements which would be needed in order to produce a 17-ft canal 'complete and proper for work as originally proposed'. His estimate for this additional work was £104,490; and even this did not include the cost of establishing steam-tugs on the lochs, which he considered to be now essential if the canal was to operate efficiently and profitably. On the suggestion that navigation should be discontinued, Walker reminded the Treasury that this would not significantly reduce the cost of maintaining the works; and he pointed out that abandonment would be as expensive as reconstruction.[43]

In his evidence to the 1839 Select Committee Walker stressed the need for a second lock at Gairlochy, pointing out that a failure of the existing lock would release a volume of water 20 ft deep and 6,000 acres in area. George May reported that 'considerable dilapidations' had taken place at Fort Augustus since the failure of the bottom lock, and that temporary repairs had also been necessary at Banavie and in the Laggan cut; but he endorsed Walker's view that a sum of £200,000 would be 'perfectly sufficient' to put the canal in good order, 'even supposing adverse circumstances'. Other witnesses reported the views of merchants and shipowners in Glasgow, Dundee, Belfast and London. The chairman of the Glasgow Chamber of Commerce maintained that an improvement of the canal would facilitate trade between the Clyde and the Baltic, and the director of the Dundee, Perth & London Shipping Company believed that trade between the Tay and the Irish Sea would use a reconstructed Caledonian in preference to the Forth & Clyde in order to avoid the need for transhipment. A Belfast merchant, while agreeing that vessels bringing flax-seed from Riga to Belfast often had a rough passage through the Pentland Firth, wondered whether they would use the canal regularly even if it were 'perfectly available and in good order'; but two other witnesses were confident that there was no basis for such doubts.[44]

In their report, the Select Committee said that the condition of the canal called for the immediate attention of Parliament 'not only with reference to the preservation of the works but also as regards the security of life and property in portions of the districts through which the canal passes'. They were opposed to the abandonment or closing of the canal, and believed that its unfinished state had 'prevented the development of those benefits to commerce which might reasonably have been expected to result from its construction'. Being convinced that the completion and deepening of the canal and the provision of steam-tugs would offer considerable advantages to 'the trade of several of the large and important northern sea-ports', they recommended that the government should make a grant of £200,000 for these purposes.[45]

In March 1840 the Select Committee was reappointed to consider proposals for the completion of the canal by private enterprise; and when they reported again in June they approved this plan but added that if it could not be put into effect within six months the repairs should be undertaken by the government and the Commissioners. An Act of August 1840 authorized the leasing of the canal free of rent to a joint stock company; but no offers were received, and the Commissioners therefore remained in control. In 1841 Parliament voted £25,000 towards the debt contracted by the Commissioners for urgent repairs, and the Treasury decided to send Sir Edward Parry, the arctic explorer, to assess the commercial benefits to be expected from the canal's completion.[46]

Having inspected the canal by boat and on foot and consulted shipowners in Aberdeen, Dundee, Leith, Glasgow, Newcastle,

Hull and Liverpool, Parry reported in March 1842 that it would be possible, if horses and steamboats were available for towing on the canal and the lochs respectively, to reduce the time needed for navigation to 72 hours in winter and 42 in summer. At the same time he recommended that buoys and beacons should be placed at the approaches to the canal, and that steam-tugs should be established both on the Inverness Firth and on Loch Linnhe. While recognizing that the Caledonian was subject to competition from the Forth & Clyde, the new Edinburgh & Glasgow Railway, the Newcastle & Carlisle Railway and the railways between Hull and Liverpool, he was confident that 'no great quantity of heavy goods would be conveyed by railway' and that merchants would gladly use the larger in preference to the smaller canal to avoid transhipment. Like most of the shipowners he consulted, he failed to realize that the development of steam navigation was not only making it possible to overcome the problems of the canal's lochs but also reducing the terrors of the Pentland Firth.[47]

George May, giving evidence in April 1842 before the Select Committee which had been appointed to examine Parry's report, explained that there had been further masonry failures at Corpach and Banavie in 1840, and that the funds for essential repairs had been borrowed from the Bank of Scotland. As before, he laid stress on the danger of flooding from Loch Lochy; and he told the committee that the fear of this had sometimes made people 'take to the hills to escape'. James Walker reiterated his view that the whole canal should be put in working order; and the committee concluded that while no adequate return on expenditure could now be expected it would be a grave mistake to abandon such a 'great national work'.[48]

The government made a grant of £50,000 in July 1842, and the Commissioners asked Walker and his partner W. Burgess to prepare detailed plans for the repair and completion of the canal. Part of the masonry of the Gairlochy lock had to be rebuilt in January 1843, and two months later the arch of the Upper Banavie culvert collapsed; after examining the culvert, May declared himself 'thoroughly impressed with the difficulty, expense and delay that would be involved in rebuilding it', but observed that such an accident might have had 'infinitely more fatal results'. Having received the plans of Walker and Burgess, the Commissioners entered into a contract with Jackson and Bean of Birmingham for the work required; and the operation, which was now expected to take three years and cost £136,089, began on 25 September 1843. Its urgency was further demonstrated the following April, when

Bean fell with his horse from an insecure bridge and was killed; and it was belatedly authorized by Parliament in October.[49]

The only landowner to protest was Evan Baillie of Dochfour, who objected to the proposed embankment through Loch Dochfour near his house; May dismissed his objection as a 'mere matter of taste', and declared that the embankment could be made 'ornamental'. The first stone of the additional lock at Gairlochy was laid by James Loch in October 1844 'in the presence of a numerous assemblage, including many of the principal heritors of the county'. Reporting on progress in February 1845, Walker and Burgess said that the construction of a proper passing-place in the middle of the Banavie locks had been found so difficult that it had been abandoned, but that they hoped to expedite passage through the locks by improving the sluices and the gate-machinery. The number of men at work in the quarries and on the canal was now between 1,000 and 1,500.[50]

Reporting again in June 1846, Walker and Burgess referred to the rise in prices since the commencement of the work, and explained that it had proved necessary to construct three extra culverts under the Banavie reach, erect new bridges 'for the convenience of Mr Baillie', and widen the Dochfour embankment so that it would carry a road. The consequence of these factors was an increase of £16,500 in the cost of the improvements. Steps were taken in 1846 for the lighting of the canal entrances and for the placing of buoys to mark the channel between Kessock Ferry and Fort George; and in 1847 the Commissioners ordered four tugs, which were to be based on Loch Linnhe, Loch Lochy, Loch Ness and the Inverness Firth. These vessels were named the *Speaker*, the *Secretary*, the *Hero* and the *Engineer*, and cost £3,200, £2,500, £2,000 and £1,800 respectively; the first two had 50 hp and the other two 40 hp engines. The canal was reopened for through traffic on 1 May 1847; it now had 29 locks, and was 17 ft deep throughout. The government grants for the reconstruction and for the purchase of the tugs had totalled £228,000.[51]

In October 1845 a private company sought permission to make a railway along the south side of the canal; but the Commissioners 'entirely discouraged the idea, at all events until the existing rage for railways should have passed'. Plans were made for the establishment of new wharfs above the locks and bridge at Muirtown; and by an Act of July 1847 the Commissioners and the Inverness Harbour Trustees were empowered to charge shore dues on goods loaded and unloaded in the basin below the bridge and the Commissioners and the Town Council were given the right to levy petty customs on goods landed at the new wharfs above the bridge.[52]

1846–7 was a year of severe distress in the Highlands because of the failure of the potato crop. The government expressed the hope that Highlanders employed on the canal would not be dismissed, and that work would be given to all labourers who applied for it; and the Commissioners therefore took on a labour-force in excess of their requirements, especially in the western district where the situation was most serious. In the autumn of 1847 Prince Albert visited the canal, travelling in a screw-steamer of 317 tons named the *Fairy*. Some alarm was caused by the fact that the *Fairy*'s propeller was unable to check the boat effectively when it approached the locks; and the Commissioners were prompted by this experience to withdraw the two-thirds reduction in tonnage dues which they had previously offered to screw-steamers. In the following year Cameron of Lochiel opened a new inn at Banavie, the Commissioners having closed the existing one and undertaken not to sanction another.[53]

In 1848 a timber jetty was built on the north-west wing-wall of the Corpach sea-lock, and a ferry-pier was established at Bona on the bridgeless and lockless channel between Loch Dochfour and Loch Ness. Tonnage dues on unladen sailing-vessels were fixed at 1s per ton for a complete passage or ¼d per ton per mile for any distance less than 48 miles; tonnage dues on laden sailing-vessels were 25 per cent, toll dues on goods 50 per cent, and tonnage dues on unladen and laden steam-vessels 100 per cent higher. Passenger steamers on the Glasgow–Inverness service were now making from two to five passages of the canal per day. The charges on these steamers were said to be low enough to 'bring the opportunity of visiting the Highland districts within the reach of thousands' who would not otherwise have had this chance; and it was claimed that the Crinan and Caledonian Canals were 'thus almost supplying the place of railroads in a country from which the natural features and the scarceness of the existing population had as yet excluded them'.[54]

January 1849 was stormy, and there was serious flooding in Inverness. The canal bank was breached at Dochgarroch, at Aberchalder, and in the reach between Aberchalder and Kytra; and flood-water from Loch Ness poured through the breach at Dochgarroch until the water-level in the loch was 6 ft below the regulating-weir. Walker reported that the canal could be restored to its former condition for £10,000; and in March this sum was granted by Parliament. As a result of the flooding the Commissioners were threatened with claims for damages. J. Inglis Nicol of the Ness Woollen Manufactory claimed £5,220; and the inhabitants of Inverness complained that an outbreak of cholera in the town had been caused by water from the canal. Nicol's claim, however, was annotated in the minutes with a double exclamation-mark; and the Commissioners told May that they had not admitted liability for any of the damage done by the flood-water. In April 1850 Nicol received £500 in compensation; and in July 1854 the Inspector of the Poor for Inverness was given permission to draw water from the canal in order to 'flush the drains of the west part of the town', the fear of cholera having rendered this measure 'advisable'.[55]

The repairs necessitated by the flood were completed by July 1850; and in order to prevent a recurrence of such damage the south-east bank of the canal above Dochgarroch was raised to 2½ ft above the highest level of the flood-water, the banks of the reach above Aberchalder were raised by 2 ft, and the banks above the Laggan locks were raised by 3 ft. A house was built at Dochgarroch later in the year for the lock-keeper; and four 'substantial' houses were provided for employees in Fort Augustus. Dues on screw-steamers were reduced again to 1s 6d per register ton for the complete passage, the dues on other steamers remaining at 2s. Hutcheson's took over the passenger-service in 1851, and in the following years travellers from Inverness to the south were carried by the *Curlew* and the *Edinburgh Castle* from Inverness to Banavie and by the *Dolphin* from Fort William to Oban, where there was a connection for Crinan.[56]

There were frequent complaints at this period about the use of the canal on Sunday. In July 1850, for example, the Moderator of the Free Church Presbytery of Abertarff asked the Commissioners to put a stop to all 'unnecessary labour' on Sunday, and protested against the regulation whereby tugs on special business were to be passed through the locks at all times. The Commissioners, however, refused to make any change. Another contentious issue was the alleged collaboration of the lock-keepers with local poachers. About 1850 Evan Baillie of Dochfour drew the attention of the Speaker, who was chairman of the Commissioners, to a case in which an employee of the canal had 'by his own confession been implicated in the clandestine conveyance of stolen venison on the canal'. The Speaker ordered the culprit to be dismissed, and gave a warning that all similar charges would be closely investigated and 'if substantial would be severely punished'. In June 1852, however, Baillie remained 'entirely persuaded that the same practice still prevailed, and that almost every lock-keeper on the canal was more or less implicated in poaching or receiving poached game, and that very stringent rules would be necessary to put it down'.[57]

Hopes for a dramatic improvement in trade after the reconstruction and the introduction of tugs were not fulfilled. Explanations offered included the increased rail communication between the east and west coasts, the reduction of grain imports from the continent, and the increasing size of the vessels engaged in trade with the Baltic. In 1852–3 revenue dropped from £7,909 to £5,888; this fall was attributed in part to the less frequent use of the canal by the Moray Firth herring-boats and in part to a bad potato-harvest in the northern counties, which had been sending potatoes in increasing quantities to Ireland and Liverpool. The failure of the Irish potato-harvest in the following year stimulated the canal's southward trade in seed potatoes; but the herring-boats were increasingly diverted to the Baltic, the Irish market for herring having been destroyed by famine and emigration. The higher freight costs during the Crimean War increased the price of coal in the north-west and thus made the running of the canal's tug-boats more expensive. When a lock-keeper of forty years' service was drowned in a lock, the Commissioners decided because of the 'very precarious' state of their finances not to assume any 'permanent burden' for his widow and child but to give them 'a gratuity of £10' when they left their house.[58]

An Act of August 1857 empowered the Commissioners to erect piers along the line of the canal and levy pier dues in order to defray the expense. The canal's commercial position was investigated by the Treasury in the same year, and it was recommended that an ice-breaker should be acquired so that stoppages for frost could be avoided in future. Revenue was now suffering from the fact that the Pentland route had been made less perilous by the preparation of new charts and the erection of numerous lighthouses, beacons and buoys; and in 1858 it was reported that the spread of potato-disease in the north was reducing not only ordinary revenue but also revenue from the tugs, which had been regularly employed by vessels trading in potatoes. Because of the spread of steam navigation, the tugs were proving less remunerative than had been hoped; and after a vain attempt to let them to private owners the Commissioners resolved to sell two of them and use the proceeds to repair the other two.[59]

In an attempt to increase revenue the Commissioners decided in 1860 to charge transit dues of 4d to 1s on all passengers travelling from either end of the canal to beyond Fort Augustus; and by an Act passed in July of that year they were given the right to charge wharfage dues of 6d per ton on all vessels using the canal basins. The same Act fixed toll dues on goods at 3d per ton per mile, tonnage dues on laden vessels at 2d per ton per mile, tonnage dues on unladen vessels at 1d per ton per mile, and pier dues at 1s for horses and four-wheeled carriages, 6d for two-wheeled carriages, 3d for sporting dogs, 2d for passengers and loaded wheelbarrows and 1d for calves. The Commissioners were again given power to lease the canal free of rent; but offers can scarcely have been expected. The annual deficit was said in 1863 to be about £1,000; and this was explained by reference to the depressed state of the coasting trade and the Moray Firth herring-fishery, the declining traffic in home-grown grain, and the effect of bad weather on the revenue from passengers.[60]

In 1864 the wooden pier at Bona Ferry was replaced by a stone one. Local traffic remained unprofitable; but the American Civil War caused a slight increase in the flax and linseed trade from the Baltic. The Perth–Inverness railway was completed in 1865, but the Commissioners were perturbed by this only because it meant the end of a profitable iron trade; the railways regarded as serious competitors were the coast-to-coast railways further south, which encouraged Glasgow and Lancashire to trade with northern Europe not directly but through the ports on the Forth and the Humber. Improvements under way included the erection of new houses for employees and the dredging of the navigable channel in Loch Oich; 15,560 passengers travelled through the canal in 'fast and slow steamers' in the summer of 1863; and three years later the steamboat *Gondolier*, which Victoria was to use on her passage through the canal in 1873, came into operation on the Inverness–Banavie service. The total number of passages by sailing and steam vessels rose from 1,695 in 1865 to 2,004 in 1866.[61]

There was flooding from the River Ness in February 1868; and though no damage was done to the canal the Inverness Town Council claimed that the floods had resulted from the Commissioners' having altered the outlet of Loch Ness. An inquiry was ordered, and Sir John Fowler reported that the canal works had done nothing to promote flooding but that the restricting and obstruction of the river-channel through Inverness by roadways and a water-main had contributed to the raising of the water-level. The timber growing along the canal between Banavie and Gairlochy was felled in 1870, and arrangements were made for its replacement. A number of accidents having occurred at Gairlochy where horses and carriages gathered to meet the swift steamers, it was decided in 1873 that a timber palisade should be erected there; and in 1875 a new steamer-wharf was completed at Muirtown. In the following year the Highland Railway Company were given land for a line from Muirtown to Inverness, and a new tug to replace the *Speaker* was ordered from Cunliffe and Dunlop at a cost of £2,150. By 1880, this tug was the only one in service.[62]

Revenue fell from £8,254 in 1877–8 to £7,355 in 1878–9, the drop being attributed to 'the general depression of trade'. A debt of £5,284 was due to the Bank of Scotland; and the need for repairs, particularly to the Corpach, Banavie and Gairlochy locks, caused serious concern. In 1881 the Commissioners received a grant of £10,000, which enabled them to discharge their debts and undertake 'extraordinary' repairs. Further repairs became necessary in the following year when the screw-steamer *Rockabill* ran into one of the lock-gates at Fort Augustus; the canal had to be closed for eight days, but the vessel's owners, the Clyde Shipping Company, admitted liability. In 1883, the Glasgow & North-Western Railway Company brought in a bill for a railway along the entire length of the canal; but this proposal was defeated in Parliament.[63]

George May died in August 1867 and was succeeded first by James Davidson, then in 1878 by William Rhodes, and then in 1885 by John Davidson, who had been resident engineer of the Crinan. Reporting on the state of the lock-gates in June 1888, Davidson said that 27½ of the 42 pairs were of cast iron with a sheathing of timber, while 12½ pairs were of timber only and 2 pairs were of oak, larch and steel with a sheathing of pitch-pine. The cast-iron gates were in an unsatisfactory state, and the masonry of several locks had suffered from their excessive weight; and since he considered cast iron too brittle a material for an age of heavy steamers, Davidson asked the Chancellor of the Exchequer for a grant of £20,000 so that all the lock-gates on the canal could be replaced by new ones of oak and steel. This sum was granted in instalments from 1890 to 1893, and by 1906 all the gates had been renewed.[64]

Passenger traffic remained important despite the existence of a rail link from Inverness to the south. There was a period of fierce competition, later exaggerated in the legend that one company had cut the Glasgow–Inverness fare to 6d and another had promptly made it 'nothing and a bottle of porter thrown in'; but in due course MacBrayne's gained control of the whole passenger trade. In 1893, they were operating 'commodious' steamers between Inverness and Oban; and the Inverness guide-book advised tourists to catch the 7 am boat from Muirtown to Fort Augustus, spend 3¼ hours enjoying 'the beautiful surroundings of the village', and then join the north-bound boat at 2.15 pm for the return journey. The opening of the West Highland Railway's Banavie branch in 1895 brought a 'perceptible increase' in the number of passengers on the canal; and the construction of the Invergarry & Fort Augustus Railway, which was authorized in 1896 and finished in 1903, provided a brief stimulus to goods traffic.[65]

The general opinion of those who gave evidence to the Royal Commission of 1906–9 was that the canal was 'antiquated' and 'practically useless' even for ordinary coasters. The only industrial establishment on the line was the new British Aluminium Company works at Foyers; and it was alleged that the development of industry had been handicapped by the fact that the locks would not admit large steamers. Besides criticizing the smallness of the locks, which were said to confine navigation to vessels specially designed for them, witnesses objected to the canal's shallowness and sharp curves, to the problems created by tidal conditions at the sea-locks, to the fact that the locks were manually operated, and to the frequently inadequate water-level in Loch Oich. The Inverness County Council contended that an enlarged canal with fewer locks would attract new industries to the area and thus help to prevent emigration. A civil engineer recommended lowering Loch Oich to the level of Loch Lochy and removing all the locks between Banavie and Fort Augustus; and an Inverness JP suggested that the locks should be lengthened and that the depth should be increased to 30 ft. The technical adviser to the British Aluminium Company explained that they had chartered a vessel to bring materials to Foyers from Glasgow and Ireland, and that they would like to see both the Caledonian and the Crinan made wider and deeper.[66]

In his evidence to the Commission, Davidson said that the depth had 'of late years' been reduced to 15 ft because sand used to stop leaks had raised the level of the canal-bed. The revenue for 1904–5 had been £8,102; and this had included £1,472 in tonnage dues on passenger- and goods-steamers, £1,436 in tonnage dues on fishing-boats, £1,396 in tonnage dues on cargo-steamers, £1,309 in transit dues on passengers, and £699 in tonnage dues on passenger-steamers, besides smaller sums for rents, wharf dues, shore dues, customs dues, towing charges and tonnage dues on tugs, yachts and dredgers. The passenger and goods steamers, Davidson explained, operated between Inverness and Glasgow, whereas the passenger steamers went from Inverness to Banavie; from Banavie, passengers were taken by rail to Fort William, and then by steamer to Oban and through the Crinan Canal to the Clyde. Davidson was against the enlargement of the canal, believing that the larger vessels would still use the Pentland route to avoid dues; and in their final report the Commission took the same view.[67]

One of the Laggan locks collapsed in January 1910 and had to be repaired at a cost of £155; and soon afterwards serious defects were reported in the original masonry at Corpach, Banavie and Fort Augustus. Davidson resigned in 1912, and L. John Groves assumed responsibility for the Caledonian as well as the Crinan. Revenue declined from £10,686 in 1911–12 to £9,770 in 1912–13, largely because of a fall in the number of passages by fishing-boats. During the war the passenger-service was curtailed, and revenue was further reduced by the fact that no dues were paid by the many fishing-boats and cargo-steamers which had been re-quisitioned by the Admiralty; but these losses were partly made up by Treasury loans totalling £20,000 in the years 1916–19. In the latter part of the war American naval bases were set up at Muirtown and on the Cromarty Firth to help in laying the mine-field from Orkney to Norway which was to exclude German sub-marines from the Atlantic; and 48,000 tons of mines and naval stores were shipped to these bases through the canal.[68]

An Act of August 1919 transferred control of the canal to the newly-established Ministry of Transport; and in the following year the Commissioners made their final report, attributing the canal's failure to regain its pre-war goods and passenger traffic to unfair competition from the railways. Revenue for 1919–20 was £9,036; but expenditure was £22,360, the increase being caused by the rising cost of labour and materials. The Ministry of Transport made a sum of £11,000 available in 1920 so that a general repair of the masonry could be carried out; the canal was closed for nine weeks, and the eight locks at Banavie were thoroughly restored. No action was taken, however, on Groves' proposal for making Loch Ness the summit, reducing the number of locks to 6, and increasing their size to 500 ft by 60 ft, or on his alternative pro-posal for making Loch Lochy the summit, reducing the number of locks to 10, and increasing their size to 350 ft by 55 ft.[69]

Commenting on the anniversary of the opening in 1922, the *Glasgow Herald* observed that it was 'questionable whether the canal had ever justified the expense it entailed'; and a few days later the Ministry of Transport declined to consider plans for its reconstruction. In 1927 the MP for Inverness drew the government's attention to the inadequacy of the canal's bridges, which were too weak to bear fully-loaded buses. In 1929 a drifter burst through two lock-gates at Banavie, causing flood-damage to the value of £4,000, and the canal had to be closed for three months for repairs. The residual passenger-service between Inverness and Fort Augustus was withdrawn in the same year; but pleasure-trips on Loch Ness were maintained until the outbreak of war. Tolls were reduced in 1937 to 1s 6d per ton on dredgers and laden oil-tankers, 2s 6d per ton on other vessels carrying dangerous goods, 1s 3d per ton on tugs, and 1s per ton on other vessels; the charge for a canal tug was fixed at £6 per day. Proposals for widening the canal were put forward by Inverness Town Council in 1931 and 1944, but received little support. In the Second World War the canal was again used for the carriage of military goods; and the number of through passages rose from 726 in 1939 to 1,107 in 1946.[70]

The British Transport Commission, reporting in 1955 on the canals they had acquired seven years earlier, expressed the hope that traffic on the Caledonian might be stimulated by afforestation, the development of hydro-electric power, and the establishment of an atomic station at Dounreay in Caithness. The 1958 committee of inquiry, however, reported that it was impracticable for the Forestry Commission to use the canal because the destinations to which timber was sent were not directly accessible by boat, and that the light industries which were expected to follow the hydro-electric and atomic schemes might prefer road to water transport as the aluminium works and the farms in the glen had come to do. Steamer-trips on the canal having been discontinued, it was now used chiefly by fishing-boats, which made up 65 per cent of the total, and by small cargo-vessels designed for the inland waterways of the continent. Users of the canal found that they saved fuel, but that the slowness of the hand-operated locks and the lack of illumination at night made a journey through the Caledonian as time-consuming as one by the Pentland Firth.[71]

In 1960 the Commission initiated a scheme for the mechanization of all the locks on the canal so that vessels could pass through them more quickly; and in September 1963 the vice-chairman of the British Waterways Board, which had assumed responsibility for the canal in 1962, inaugurated the newly-mechanized flight of four at Muirtown, expressing the hope that 13 of the 29 locks would be mechanized by the end of the year. In 1964 the Board signed a 30-year agreement with Wiggins, Teape & Company for the enlargement of the basin at Corpach to receive up to 100,000

XIX. Caledonian Canal: (*above*) the Laggan cut, looking north; (*below*) Invergarry Castle and Loch Oich about 1830

XX. Caledonian Canal: (*above*) the abbey and locks at Fort Augustus; (*below*) Muirtown Basin in 1820

tons a year of raw material for the £20,000,000 pulp-mill the company was building there. The canal was closed for ten months so that the sea-lock and the basin could be adapted to receive vessels of 1,000 tons; and when it reopened in April 1965 there were two wharfs, one being for the pulp-mill and the other for general trade. The Corpach basin in its new form can accommodate vessels 203 ft long and 35 ft wide, and the whole canal can take vessels 150 ft long and 35 ft wide; the maximum draught in each case is 13½ ft.[72]

Canal dues were reduced in November 1965, and it was claimed that this reduction would make the canal 'almost always' cheaper than the stormy passage round the north coast. The pulp-mill at Corpach was opened in 1966, and the southern end of the canal has since become a busy centre of industry with a rapidly-expanding population. Despite the renewed activity in the terminal basin, the canal as a whole is now used chiefly by fishing-boats and yachts; but there is a limited trade in such commodities as grain, oil, coal, salt, chemicals, fertilizers and building materials, and the pleasure-cruises from Muirtown Basin have been given a new lease of life by curiosity about the Loch Ness Monster. All the locks have now been mechanized except the two at Gairlochy; and it is hoped that these will be dealt with in 1968.[73]

CHAPTER VIII

Minor and Proposed Canals

FOR reasons which are immediately apparent on a contour map, most of the minor canals built or planned in Scotland were conceived not as contributory parts of a waterway system but as outlets from centres of industrial and agricultural production to the most convenient starting-points for loch, firth or sea navigation. While few of the Scottish rivers were found suitable for canalization, their valleys seemed for both topographical and economic reasons to offer the most eligible routes for artificial waterways; and the mountain-masses of the Highlands and the Southern Uplands restricted the opportunities for canal-construction across the watersheds dividing these valleys. Schemes were put forward for the development of a network based on the Forth & Clyde, and for the carrying of canals from the Clyde, the Tay and the Angus coast into the line of valleys which links Dumbarton with Stonehaven; but despite the boldness of the sea-to-sea schemes the characteristic project of the Scottish canal-promoter was an isolated waterway offering inexpensive carriage to the nearest harbour.

In the south-west, the valleys of the Annan, the Nith, the Dee and the Water of Fleet were obvious canal-routes, and plans were made for all four. The Annandale Canal, which was proposed about 1810 by a civil engineer named Jardine, was to draw its water-supply from the Annan at Kirkbank, and was to run 'along the hollow on the east side of the town of Lochmaben' and past 'the lime quarries of Kelhead' to enter the Annan 'where the tide rises to a considerable height at the Old Mill harbour'. The *Dumfries Courier and Herald* gave the depth of the intended canal as 4½ ft, and offered a poem in its praise:

> Already fancy sees the line complete,
> Which Solway knits to Bruce's ancient seat,
> And as the eager crowd exulting gaze,
> Marks in each varying mind thought's devious maze.

The aged matron not inclined to roam
Sees busy barges gliding near her home,
And—present things contrasting with the past—
Wonders to what the world will come at last.
The shepherd boy spread on the distant heath
Unwearied eyes the living scene beneath;
Sums up the vessels, and will often try
On each extreme another to descry.

The matron and the shepherd were doomed, however, to be disappointed.[1]

The plan for a Nithsdale Canal originated in the desire of the magistrates of Dumfries to improve the navigation of the river. Smeaton examined the channel, and reported that it would be impracticable to provide a satisfactory waterway 'otherwise than by cutting a navigable canal with proper locks'. Consideration was given to the possibility of making a canal from the Nith near Dalswinton by way of Tinwald, Torthorwald and Lochar Moss to the Solway at Caerlaverock Castle; but the scheme was given up 'because of lack of funds'. Some improvement in the river-navigation was finally effected in 1811, under the direction of James Hollinsworth.[2]

In the valley of the Dee two short canals were made between 1765 and 1789 by Sir Alexander Gordon of Culvennan. The first, which was known as Carlingwark Lane, ran from Carlingwark Loch (which had immense beds of shell-marl) to the nearest point on the river; and the second left the river west of Old Greenlaw and rejoined it below Glenlochar Bridge. In conjunction with the river and its lochs, they provided a through communication between Castle Douglas and New Galloway and were used during their short lifetime for taking marl to the farms on this route; the marl was carried in flat-bottomed boats, the largest of which could take 400 cu ft. A more ambitious scheme for a canal to serve the whole valley was authorized by the Glenkens Canal Act of 1802. This canal was to have run from 'the boat-pool of Dalry in the Glenkens' along the east side of the Rivers Ken and Dee until it joined the tidal reach of the latter near 'the port and town of Kirkcudbright'. It was hoped that it would 'facilitate and render less expensive the conveyance of manure, coal, lime and all sorts of commodities'; and the company was given the power to raise a capital of £30,000 in £100 shares. After surveying the route, however, Rennie estimated the cost of construction at £33,382; and in view of the limited character of the probable trade it was decided that so large an expenditure would not be justified.[3]

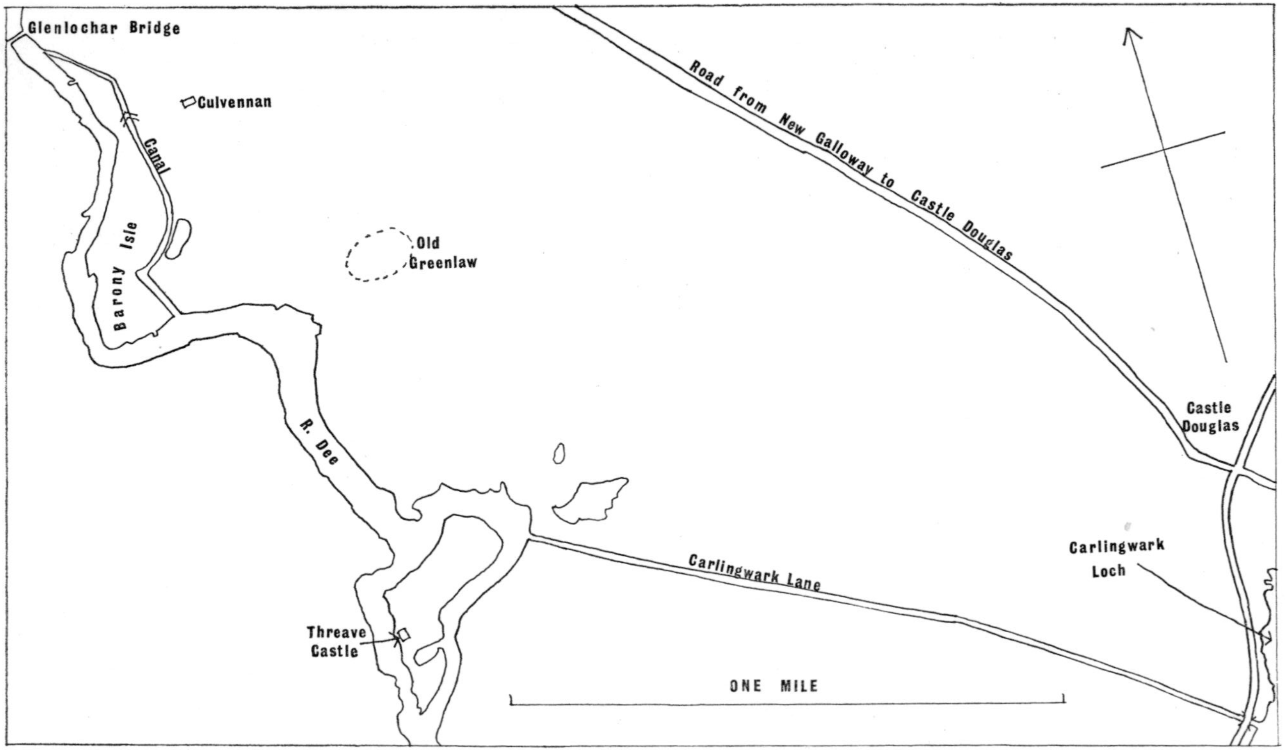

14. Carlingwark Canal

The Water of Fleet between Gatehouse and the sea had been navigable before the canal era; but a small canal scheme was carried into effect in 1824 in order to improve the navigation and reclaim land previously covered at high tide. Alexander Murray of Broughton, the owner of Cally Park on the eastern shore of the estuary, brought men from his estates in Donegal so that they might pay off their arrears of rent by digging a new cut for the river. Work began on 17 June and finished on 3 October; and the total cost of the 1,400-yd navigable channel was £2,204.[4]

Further north, in Kyle and Renfrewshire, the promotion of canals was stimulated by the needs rather of industrial than of agricultural development. In the 'desert and inland' parish of Muirkirk, for example, an iron works was set up in 1789 to take advantage of the cheap raw materials available; and a 1-mile canal was dug to 'supply the machineries with water' and facilitate the transporting of ore and coal to the furnaces. A 1792 letter from J. L. McAdam to an Ayrshire landowner mentions plans for a coal canal from Muirkirk to Ayr; but this was never made.[5]

As the carpet industry expanded in Kilmarnock, the need for better transport made itself felt; and in 1797 proposals were advanced for a new harbour at 'the Point of Troon' and for a canal from there to the western outskirts of Kilmarnock by way of Hillhouse, Auchans and Fairlie, with a branch from Fairlie along the south bank of the Irvine to Caprington and Riccarton Bridge. No immediate action was taken, however; and in 1806 the Marquis of Titchfield, who owned land at both ends of the route, declared his preference for a railway. In a letter to the Earl of Eglinton, he wrote:

> This preference has arisen from the consideration that the adoption of it will be attended with much less prejudice to the land through which it may pass, and that, as it can be completed at much less expense than a canal, it will eventually allow the business of the country to be done on much cheaper terms.

The railway, which followed a more northerly course than that which had been chosen for the canal, was authorized in 1808, and opened in 1812.[6]

In the parish of Stevenston, coal was mined for export to Ireland from before 1700; but land carriage to the coast was expensive, and when Robert Reid Cunningham inherited the estate in 1770 he built a 2¼-mile lockless canal from the mines to a point within 600 yd of Saltcoats harbour. This canal, which was finished and navigated on 19 September 1772, had a minimum width of 12 ft at bottom and a minimum depth of 4 ft; and it was 'very wide and deep in some places from the inequality of the ground'. As new pits were opened along the route, 'long side-branches' were cut to serve them.[7]

Another coal-canal projected in the 1770s was a 3-mile cut from the colliery of Hurlet to the town of Paisley, for which James Watt made a survey in 1773. Watt's estimate was only £4,600; but his plan was never put into effect.[8]

Both before and after the building of the Glasgow, Paisley & Johnstone Canal there were attempts to provide better transport facilities for Paisley's cotton-mills by straightening and deepening the White Cart. An Act for this purpose was passed in 1753; and in 1787, when a new road-bridge at Inchinnan had rendered some improvement essential and Golborne's work on the Clyde had created both a model and an additional incentive, the Paisley Town Council obtained a further Act 'to improve the navigation of the River Cart and to make a navigable cut or canal across the turnpike road leading from Glasgow to Greenock'. The digging of this 'navigable cut', which was a new course for part of the river, began on 23 August; and it was expected that the work would be completed the following year. Another scheme for improving the river and 'extending and deepening the harbour of Paisley' was initiated in 1835 in anticipation of the Forth & Cart Canal Act; and in the 1880s a scheme was prepared for making the Cart navigable for ocean-going vessels. The river was declared ready for this purpose on 25 May 1891; but on the same day the *Joseph* ran aground at the harbour entrance, and in the following years the plan was abandoned.[9]

In conjunction with the Cart Navigation, the Forth & Cart was designed to serve as a feeder from Paisley to the Forth & Clyde. Its promoters, who were reviving a scheme considered by Hugh Baird in 1799, hoped that it would provide an uninterrupted communication between Paisley and the Firth of Forth which would be preferred to that via Port Eglinton and Port Dundas, and that coal from the Coatbridge district would be brought to Paisley by way of the Monkland Canal, the Forth & Clyde, the new cut and the River Cart; and they claimed that their scheme would reduce the consumption of water on the Forth & Clyde by providing a link to the Clyde for 'boats and barges of small burden' which would otherwise have had to pass through the larger locks to the west. The canal, which was opened in March 1840 was ½ mile long, and ran from the Clyde opposite the mouth of the Cart to the Forth & Clyde at Whitecrook; and its level was raised 30 ft by 1 single lock and 1 staircase pair, each chamber being 67 ft long and

15 ft wide. An Act authorizing its acquisition by the Forth & Clyde was passed in 1842; but the actual transfer did not take place until 1855. At that date, it was reported that the annual revenue averaged £325, that the annual expense of maintenance, management and interest on debt averaged £342, and that it would cost £3,100 to put the works in good order. The mineral trade to Paisley having been taken over by the railways from Ayrshire, the annual traffic was less than 40,000 tons; and the company was heavily in debt and had paid no dividend. The Forth & Clyde, which drew an average revenue of £739 annually from 'the whole Cart Canal trade', agreed to buy the canal for £6,400, and to pay the original proprietors 1d per ton on all future traffic above 90,000 tons per annum; and the Forth & Cart committee, while realizing that the cash payment would be wholly absorbed by the company's debt and that a doubling of trade was improbable, advised their shareholders to accept. The canal passed with the Forth & Clyde into the ownership of the Caledonian Railway Company; and it was finally closed under an Act of 1893.[10]

Another projected feeder for the Forth & Clyde was the Campsie Canal, which was to have run from the village and alum works of Campsie to the main canal 4 miles away. Authority for this scheme was given by an Act of 1837, the Forth & Clyde having subscribed £100 towards the cost of a preliminary survey; but no further action was taken.[11]

Four years later, the Earl of Breadalbane proposed a canal through the Vale of Leven to link the terminal basin of the Forth & Clyde with Loch Lomond. James Thomson drew up plans for a £51,286 canal with a depth of 5½ ft and a surface width of 33½ ft, and reported that there would be no engineering difficulties except the aqueduct across the Leven and the regulating-lock at the northern entry. In January 1842 the interim committee recommended the scheme to proprietors of collieries in the Monklands, of the Monkland and Forth & Clyde canals, and of lands on the route and on Loch Lomond; and the Forth & Clyde agreed that it was worthy of support as a 'profitable feeder'. By 1844, however, it was clear that there was 'no possibility of the Vale of Leven Canal being entertained'.[12]

As an extension of this plan, Breadalbane put forward another for linking the north end of Loch Lomond to the south-west end of Loch Tay so as to establish a through navigation from the Clyde to Kenmore. James Leslie reported in 1838 that this would cost between £40,000 and £50,000, and that 20 to 23 locks would be needed in Glen Dochart alone. A Glasgow merchant, in a private letter of November 1841, suggested the possibility of using inclined planes in Glen Dochart and Glen Falloch, mentioned the lead-ore at Tyndrum as a potential source of revenue, and proposed that a new survey should be made with the assistance of the Vale of Leven Canal Company; but Sir Neil Menzies wrote to Breadalbane in the same month that 'more information should be obtained for procuring a railway from Loch Tay to Crianlarich', and this advice seems to have been accepted.[13]

The one canal actually constructed in Breadalbane was a ¼-mile branch from the River Falloch to a large basin near the coaching-inn at Inverarnan. This canal, which is still in existence, made it possible for the Loch Lomond steamers to penetrate 2 miles beyond the head of the loch.[14]

The waterways planned in Argyll included (besides those which were to cross Kintyre from sea to sea) a 3-mile canal across the southern end of the peninsula to convey coal to Campbeltown from the Argyll Colliery near Machrihanish Bay. The route was surveyed by Watt in 1773; and the canal was made in 1794 by the lessee of the colliery. About 40 cart-loads of coal were then taken every day to Campbeltown, where they sold for 2s 7½d per load or 7s 10½d per ton. The canal was closed in 1856, its place being taken by a light railway.[15]

At the north end of the peninsula, the valley from Loch Crinan to Loch Awe offered yet another opportunity to admit navigation from the sea to a long fresh-water loch. The authorization of the Crinan gave the idea new attractions, and in 1793 a company was formed to make a branch from the new canal to join Loch Awe at its south-west end. The distance to be covered was increased by the Crinan Company's substitution of the Daill route for the Achnashelloch; but the scheme was not forgotten, and over a period of years estimates ranging from £10,000 to £19,000 were submitted. In 1825 it was suggested that 'the fisheries of Loch Awe would prove a source of revenue', that 'the introduction of coal would induce home manufactures and stimulate enterprise', and that the canal would facilitate the shipping of timber from Loch Awe to the Clyde; but the necessary funds were not forthcoming.[16]

The country beyond the Great Glen gave few incentives to the promotion of canals. The town of Dingwall, however, was sufficiently prosperous in the early nineteenth century to undertake an improvement of the Peffrey; and in 1815–17 part of this river was diverted into an artificial channel costing £4,365. The annual revenue from the canal was only £100; and when Telford reported some years later that it would cost £800 to repair, the Town Council petitioned the Commissioners for Highland Roads and Bridges to take it over. Negotiations with the Commissioners were unsuccessful, however, and the council therefore retained control of the canal themselves.[17]

On the southern shore of the Moray Firth, the most promising field for canal-development was the Laigh of Moray. A canal from Findhorn to Forres was 'frequently spoken of' around 1811; and there was also a proposal (frustrated by high ground) for a canal from Garmouth along the course of the Spey to Rothes. The main need in this region, however, was for better communications between Elgin and the sea; and there were two proposals for satisfying this need. In each case, the Elgin basin was to be on the Earl of Findlater's property at Bishopmill to the north of the town; but one route went west to the harbour of Burghead, while the other went north through Loch Spynie to that of Lossiemouth. The second route was shorter, and was preferred on that account; but the canal was never made. The artificial watercourses which were actually constructed in this district were designed for drainage purposes only.[18]

That canals should have been planned in Moray is not surprising; but one would not have expected them to be dug in the Cairngorms. It is said, however, that there was a 'waterway system' in the Loch Morlich district 4 miles east of Aviemore, at a height of more than 1,000 ft above sea-level. This system, which was probably based on existing streams and designed for floating timber to the Spey, was succeeded by a railway-network with a 3-ft gauge; and the railway in its turn was closed in 1918.[19]

In Aberdeenshire, the desire of progressive landowners to transport fertilizer inland produced not only the major canal in the valley of the Don but also lesser schemes for canals in the valleys of the Dee and the Ugie. The former was to have drawn its water-supply from the Loch of Skene and proceeded south and then east to Aberdeen harbour; but it was never begun. The latter, which was planned in the late eighteenth century by Admiral Fergusson of Pitfour so that barges could bring shell-sand to his estate from Scotstoun of St Fergus, was to have run along the north bank of the river to reach the sea some way north of Peterhead. Part of this canal was made, and can still be seen; but there is no evidence that it was ever completed or used.[20]

'The establishment of conveyance by water to the Vale of Strathmore' was much discussed between 1760 and 1830; and many possible routes were considered. In 1767 the Earl of Findlater and Seafield declared that 'the county of Angus wanted fires', and that a better provision for the importing of coal was 'of the utmost consequence' for the prosperity of the region; and in the same year George Young paid William Keir to make a survey for a canal into Strathmore from the Tay at Perth. In 1769 the Commissioners of Forfeited Estates asked Smeaton to undertake a further survey for a canal on this line; but though he 'highly approved of the design', believing that it would be 'most beneficial' to an 'extended country' where there was an 'absolute penury of fuel', Smeaton was too busy with the Forth & Clyde to undertake the task. On his recommendation, the Commissioners asked Mackell and Brindley; and when they received refusals from them also they turned to Watt. Writing to Dr Small on 15 April 1770, Watt said that it would be possible to make a 36-mile lockless canal from Kinnoull Hill near Perth to run through Coupar Angus to Forfar; but though he proposed this to the Commissioners they took no further action. In 1788 George Dempster of Dunnichen employed Robert Whitworth to make a plan and estimate for a canal from Arbroath to within 2 miles of Forfar; and Whitworth reported that this would cost £17,788. Despite the 'want of fuel' in Angus, however, this plan was vigorously opposed by those who feared it would divert water-supplies from the local mills.[21]

In the next thirty years, there were proposals for a canal from Dumbarton via Stirling and Perth to Stonehaven, and for a 'main trunk' from Perth to Stonehaven with branches into all parts of Strathmore; George Robertson in 1810 observed that the latter scheme would require very few locks, but that the canal would be 'of no use' as no one would send goods by it. There was also a plan for linking Brechin and Montrose, the estimated cost of this being only £10,000. In 1817, however, Robert Stevenson was commissioned by the magistrates of Arbroath and Forfar to re-survey the route for a canal between these towns. In his report, Stevenson proposed a 21-lock canal 20 ft wide at bottom and 5 ft deep, to run from 'the old windmill north of Forfar' along the north side of Lochs Rescobie and Balgavies to a basin just west of Arbroath harbour. Water was to be drawn from Loch Fithie, which would be enlarged, and from a new reservoir near Myreside; and the total cost would be £88,378. Stevenson was confident that no injury would be done to the mills; and while he felt it might be 'difficult to avoid interfering with the privacy of the mansion-house', he pointed out that Dutch landowners did not object to such interference. In 1818 the proposals made by Whitworth and Stevenson were submitted to Rennie, who expressed doubts about the financial prospects for any canal terminating at Forfar and suggested that the line should be extended west along the valley. The Arbroath–Forfar scheme was taken up again six years later, the Arbroath Town Council subscribing £200 for a further report; but in 1825 it was finally rejected in favour of a railway.[22]

From the head of the Tay Navigation at Perth, canals were projected not only north-east into Strathmore but also west into Strathearn. In 1773–4, at the request of the Commissioners of Forfeited Estates, Watt 'rode over the ground' between Crieff on the Earn and Campsie Linn on the Tay. Without making a detailed survey, he reported that a canal on this line would be 'very practicable'; but he later came to the view that the 'proper line of the canal' would be down the valley of the Earn to its mouth. About 1790, it was suggested that the route might be extended inland to the east end of Loch Earn so that limestone from the 'inexhaustible' quarries on the loch could be transported to the farms in the valley; and in 1799 the various proposals were discussed in James Robertson's *General View of the Agriculture of the County of Perth*.[23]

Matters were brought to a head in 1805, when Robert Frazer submitted a new scheme to the Earl of Breadalbane, pointing out at the same time that neither he nor Rennie would accept 'any connection with the business' if Telford was employed on it. Frazer's plan, which had been prepared under Rennie's guidance, was for a new basin at the head of the South Inch, a deep cutting from there to Friarton Hole a mile downstream, and a canal from the basin to Loch Earn. Breadalbane was sufficiently impressed to take the chair at a meeting of Perthshire landowners on 16 January 1806; and at this meeting it was resolved that Frazer should make the necessary surveys 'under the direction of Mr Rennie'. Three days later Frazer wrote to Breadalbane again, suggesting that 'some hundreds of acres of good land' might be gained by lowering the level of Loch Earn; and in the spring Breadalbane sought advice from Watt about the earlier projects. In an appeal for assistance from the Commissioners of Forfeited Estates, Frazer claimed that the canal would open up a communication 'of infinite consequence to the interior parts of the Highlands', and laid special emphasis on the benefits offered to agriculture by the supply of limestone available on Loch Earn. On 3 July 1806, the House of Commons made a grant of £4,500; and in the following year the results of Frazer's survey were reported in the *Scots Magazine*. Besides proposing a basin at the South Inch, a deep cut to Friarton, and a canal 4½ ft deep from the basin via

Methven, Crieff and Comrie to Loch Earn, Frazer now suggested a branch from Methven across the Almond to Stanley and Dunkeld. By this time, however, interest in the scheme seems to have faded.[24]

The canal projects advanced in the county of Fife were less ambitious. It was suggested that 'a navigable canal upon the Eden might be carried as far up as Cupar at no great expense'; but the exports and imports of the district were judged insufficient to maintain a steady revenue. Four miles south-west of Cupar, however, there are still traces of two small canals made about 1800 to bring limestone to the top of the kilns at Burnturk Colliery and send lime and coal from the foot of the kilns to the nearby village of Kingskettle. These canals, which were 2 miles and ½ mile long respectively, were 3 ft deep and drew their water from a flooded mine; and inclined planes were used to convey material from the quarry to the beginning of the first and from the end of the second to the Kingskettle depot. The vessels used on the upper canal were of 4 to 10 tons burden, and were pulled by two men walking on opposite banks; those used on the lower canal were of 2 tons burden, and were managed from within by one man pulling on a pole suspended 3 ft above the water like a trolley-bus cable.[25]

Another Fife canal whose remains are still discernible is that said to have been made about 1490 by the Scottish admiral Sir Andrew Wood of Largo so that he could 'sail in his barge to the church every Sunday in great state'. The surviving track is in a field near Largo House, and is about ¼ mile in length; and the fact that its connection with Wood was being confidently asserted in 1792 shows at least that this was much the earliest canal in Scotland.[26]

The area round Alloa and Stirling at the head of the Firth of Forth was inevitably a centre of canal promotion. South-east of Alloa, a canal was made in the late eighteenth century to take grain from the wharfs of Kennetpans on the Forth to James Stein's distillery at Kilbagie. This cut, which was a mile long, had been obliterated by 1861; and the distillery is now a paper-mill.[27]

The main requirement in Clackmannanshire, however, was that the coal trade from the mines on the Devon should have better transport facilities than the overland carriage to Alloa on which it depended in the mid-eighteenth century. One proposal, on which Mackell and Watt made a report, was for a lockless canal from Tillicoultry to 'the school-master's house in Alloa' at a height of 30 ft above sea-level; but the scheme which attracted most attention was one for the development of the existing river-navigations, on which Lord Cathcart, the owner of collieries near Tillicoultry, sought Smeaton's advice in 1765 and 1767. In his letters to Cathcart, Smeaton discussed a plan for improving the navigation of the Forth from Alloa up to Cambus either by deepening the ford at Throsk or by making a new channel across the isthmus of Longcarse, and a plan for establishing a waterway to Cambus from just beyond Tillicoultry partly by improving the Devon and partly by artificial cuts. His estimate for the latter plan was £9,357; and he pointed out that it would be possible for another £2,408 to divert the lowest of the proposed cuts so that it finished not at Cambus but at Alloa. At the same time, he considered the possibility of carrying the canal not south to Alloa or Cambus but west either to join the Forth just above Stirling or to cross the Forth by an aqueduct near Abbey Craig and unite with another canal on the southern bank. In the event, however, the Clackmannanshire coalfield was served not by canals but by railways, the first of these being made in 1768.[28]

The canal south of the Forth with which Smeaton proposed to link the Devon Navigation was part of a scheme for improving the navigation of the Forth being put forward in 1767 by Watt. Watt's proposal at this date was that the river should be made navigable from Aberfoyle to Stirling by the establishment of locks at Cardross and Frew and the digging of a 1-mile canal from near Drip Bridge to Craigforth, and that the navigation from Stirling to Alloa should be shortened by cutting across four of the loops in the river. By 1774, however, he had extended this scheme to include Smeaton's and was proposing an improved river-navigation from Aberfoyle to Craigforth and a canal from there to Tullibody Bridge with branches to Cambus and the Devon collieries. It was hoped that this through communication from Aberfoyle to Alloa and Tillicoultry would facilitate the transportation of timber, slate and limestone from Aberfoyle to the firth and of coal from Clackmannanshire into Stirlingshire; and according to Archibald Stirling the cost was estimated by Watt at £15,000. There was also a suggestion that a branch might be taken 'through the interior part of Menteith by the water of Goodie'.[29]

In 1810, further proposals for the development of a waterway-system centred on Stirling were put forward by Alexander McGibbon in an enthusiastic book designed to 'demonstrate the advantages of small canals'. Besides proposing canals into the valleys of the Teith and the Goodie, McGibbon revived the notion of a canal from Loch Lomond to the Forth, saying that it might run by Drymen, Buchlyvie, Kippen, Gargunnock and

Stirling to Fallin, and that a branch might be made from near Fallin to Bannockburn colliery. There is no evidence that any of these suggestions was ever taken up.[30]

Twenty-five years later, when railway development was well under way, there was an attempt to link Stirling by canal to the Forth & Clyde. It was claimed that such a canal would give Stirling direct communication not only with Edinburgh and Glasgow but also (when the Forth & Cart was completed) with Paisley; and it was proposed that it should be made 4 ft deep and 25 to 30 ft wide at surface, and should run via St Ninian's, Bannockburn and Denny to join the main canal either at Lock 20 or at Castlecary. In the original prospectus the cost and annual revenue were estimated at £100,000 and £11,125 respectively; and the advantages of canals over railways were emphatically asserted, with reference both to goods traffic and to passenger traffic. After a survey of the route, John Macneill estimated the cost of a canal to Lock 20 at £90,748 and the cost of one to Castlecary at £93,681. The Forth & Clyde agreed to subscribe £50,000; but even so the subscription list was never completed.[31]

The scheme to link Bo'ness with the Forth & Clyde had a still sadder history. Having been defeated in their campaign for a small canal direct from the Broomielaw, the merchants of Bo'ness initiated a fifty years' struggle to preserve their town's position as the main port of the upper Forth. In 1768 they forced the promoters of the 'great canal' to make provision in their Act for a branch from 'the mouth of the Grange Burn' to Bo'ness. This branch was to be made by an independent company with a capital of £5,000 or £8,000 in £50 shares; it was to be 7 ft deep, and was to draw its water-supply from the River Avon 'below Jinkaboutmill'. Watt made a survey and report in 1772; but it was not until 1782 that the subscription was opened and the first general meeting held. In March 1783, the company advertised for contractors, announcing that work on the sea-lock and the western end was to start without delay. Before the end of the year, however, the £8,000 originally subscribed had been spent, and another Act had to be passed authorizing the company to raise a further £12,000 or borrow money on the security of the tolls. By this means, the aqueduct over the Avon was almost finished, and the canal was cut from there to within a mile of the town. Once again, however, funds ran out; and in 1789 a new survey was made by Whitworth, who reported that the canal would cost £17,763 to complete. At a meeting held in Glasgow in 1796 it was unanimously agreed that 'it was expedient to abandon the canal'. A customs-house was

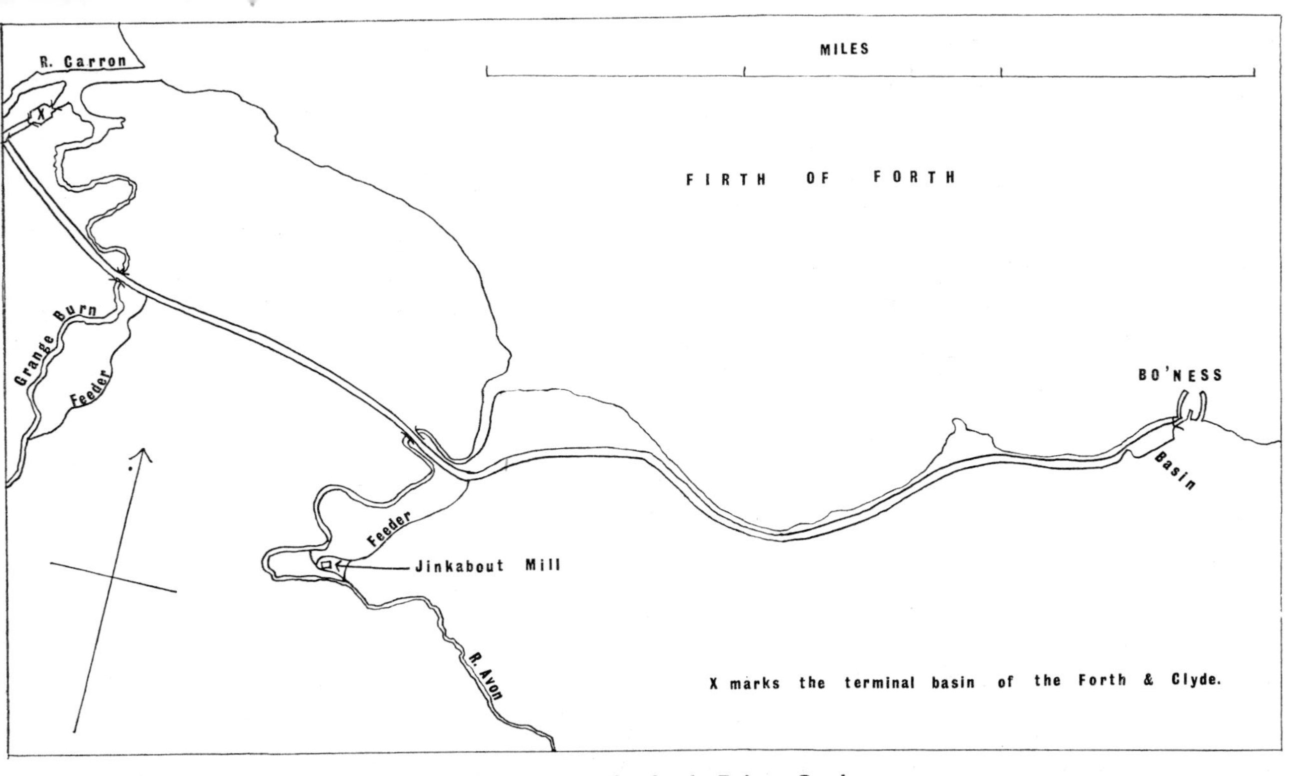

15. Whitworth's Plan for the Bo'ness Canal

established at Grangemouth in 1810; and in the next five years the receipts of the Bo'ness customs-house dropped from £30,485 to £3,835.[32]

In the south-east as in the south-west, canals were projected to link agricultural districts with the coast. Two such proposals were made in East Lothian about 1805, the first being for a 3-mile canal from the mouth of the Tyne to East Linton and the second for a 10-mile sea-to-sea canal from Tynninghame to Aberlady; but they were rejected because of the 'very limited trade of the country'.[33]

In the Merse, plans for improving inland navigation were first made about 1740, when the inhabitants of Eyemouth proposed to divert the Whiteadder into the Eye so that its trade and salmon would reach the sea at Eyemouth instead of Berwick. A company of Dutch engineers then offered to make a canal from Eyemouth via Duns to Kelso in return for a 10 years' lease of the tolls; but their offer was declined. The notion of linking Kelso with the sea was revived in 1789, when John Knox, addressing a public meeting in Jedburgh, expounded 'with his usual zeal' a plan for establishing a navigable waterway up the Tweed from Berwick. Another meeting was held at Cornhill in December, and Whitworth was commissioned to survey the country between Berwick and Ancrum Bridge. He reported, however, that a canal from Ancrum Bridge to Kelso would cost £12,047 and one from Kelso to Cornhill £14,493, and that a navigation from Cornhill to Berwick was out of the question because of the salmon-fisheries. Realizing that the trade between Cornhill and Ancrum Bridge would scarcely suffice to 'keep the canal in proper order', the subscribers decided to drop the scheme; and the *Scots Magazine*'s proposal for a canal from Kelso via Duns to Eyemouth seems not to have been considered.[34]

The successful campaign of the 1880s for a ship-canal to Manchester stimulated demands in Scotland for a comparable waterway between the Clyde and the Forth; and these demands were strengthened in the following decades by the example of the Kiel Canal and the prospect of a naval war. In 1889 an Edinburgh group commissioned D. and T. Stevenson to investigate the practicability of the scheme; and they reported that while 'a satisfactory ship-canal could not be constructed on the line of the existing navigation' it would be possible to make one from the head of Loch Long through Loch Lomond to the Forth opposite Alloa, with two locks at each end to raise ships 22 ft above sea-level. In 1890 the provisional Committee of the Forth & Clyde Ship Canal obtained a report from the Glasgow engineers Crouch and Hogg, who advocated a high-level canal 26 ft deep by the direct route from Yoker via Kirkintilloch and Kilsyth to Grangemouth. Writing in support of this plan in 1891, J. Law Crawford emphasized the strategic benefits it would afford by enabling warships to pass quickly from the North Sea to the Atlantic.[35]

The rival campaigns were continued during the period of the Royal Commission from 1906 to 1909. It was claimed that a canal by the northern route would make Stirling 'a great port'; and in support of the southern one it was said that 'Glasgow had no Clyde to the east and must have one'. The navy's need for a 'Scottish Battleship Canal' was reiterated, especially in the speeches of Vice-Admiral Sir Charles Campbell. It was said that ships at Rosyth required a 'bolt-hole' because of the risk that the Forth Bridge might be destroyed so as to block the firth; and it was suggested that Loch Lomond might be used as a naval base. The Stevenson scheme was revised by Sir John Jackson in order to meet the Admiralty's requirements; and at the Commission's request W. T. Douglas prepared an estimate for a direct-route canal at sea-level. The Commission concluded that a ship-canal through Loch Lomond would have some strategic value, but that this would not be sufficient to justify large-scale government expenditure on the project.[36]

The campaign for a Mid-Scotland Ship Canal was maintained despite the Commission's findings; and it received the support of Glasgow Town Council in 1911, of Winston Churchill (the first Lord of the Admiralty) in 1912, and of Edinburgh Town Council in 1916. By that time, despite a protest from the MP for Stirlingshire that 'we might not have another war for 100 years', attention was inevitably focused on the strategic issues. In 1917 the Admiralty obtained a new report on the rival plans from Sir W. G. Armstrong, Whitworth & Company, and resolved that 'the balance of advantage lay with the Loch Lomond route'; and it was reported in 1919 that King George V had 'for some years' been interested in the scheme, having been 'one of the first to realize its importance to the navy'. A joint committee of the Mid-Scotland Ship Canal Association and the Glasgow and Edinburgh Town Councils made representations to the Secretary of State for Scotland in favour of a canal by the direct route; but the Board of Trade, after investigating the probable use of the canal by commercial shipping, declared that they 'did not feel justified in supporting the proposal'. Sir Ian Hamilton in 1922 advocated the canal's construction as a means of providing employment for ex-servicemen; and in the following year the British Legion resolved to send a deputation on the subject to the Minister of Labour. The government resisted both this campaign and another launched by the Mid-Scotland Ship Canal Association in 1924; but in 1929 the Minister of Transport, Herbert Morrison, appointed a committee to investigate the case. While the committee was sitting, Hamilton made another speech in favour of the scheme, declaring that 'if we had possessed a Clyde–Forth Canal in 1916 Lord Kitchener would not have been drowned'. The report, however, came down firmly against the construction of the canal, saying that the Exchequer would have to bear the entire cost of about £50,000,000, that the revenue might not be sufficient to cover working expenses, and that the canal's value as a provider of employment and a stimulus to industry would be very limited.[37]

The next campaign was initiated in 1942–3 by Sir John Graham Kerr, who declared that the canal should be 'one of the first items of post-war reconstruction'. He was supported by the Scottish Unionist group in the Commons, and by a number of peers; and Sir Ian Hamilton wrote to *The Times*, observing, 'As to labour, we have many Italians available to do the bit of digging required'. Committees were set up in 1943 by the Ministry of Transport and the Scottish Council for Development and Industry; and the first of these reported in 1946, confirming the findings of the 1930 inquiry. The cost of construction was now put at £109,000,000, not including interest charges; and it was pointed out that the strategic advantages would not be sufficient to justify this immense expenditure, and that the canal would create serious problems for town planning and for road and rail communications without having any commensurate effect in providing employment. These conclusions were accepted by the government in October 1946; and when the Scottish Council committee produced its less damning report in 1947 the issue was dead.[38]

ACKNOWLEDGEMENTS

I SHOULD like to express my gratitude to the staff of the British Transport Historical Records Office in Edinburgh not only for practical assistance but also for creating an atmosphere in which it was a pleasure to work. I am also indebted to the officials of the British Waterways Board, and to the staffs of the British Museum, the House of Lords Record Office, the Scottish Record Office, the Mitchell Library, the Baillie Institute, the Waterways Museum, the Old Glasgow Museum, the Paisley Burgh Museum, the National Library of Scotland, the University Libraries in Aberdeen, Bangor, Belfast, Edinburgh and Glasgow and the Public Libraries of Aberdeen, Airdrie, Coatbridge, Dumfries, Edinburgh, Elgin, Falkirk, Inverness, Paisley and Strichen. I have to thank the Carnegie Trust for the Universities of Scotland for a grant which enabled me to travel more extensively than I could otherwise have done, and the editor of *The Journal of Transport History* for permitting me to reproduce parts of an article on the Aberdeenshire which first appeared in that periodical. Mr Charles Hadfield put at my disposal his private collection of documents on Scottish canal history, and saved me from numerous errors by his careful reading of my first draft; and Mr Jack Howdle provided much valuable information, particularly about the minor and proposed canals discussed in Chapter VIII. I am grateful, too, for the help I have received from Mr A. W. Bellringer, Mr Allan Boath, Mrs J. W. Forbes, Mrs Agnes Hastie, Mr John R. Hume, Mrs L. Keith, Mrs E. Lindsay, Mr D. McKay, Mr C. H. Scott, Mr A. Slaven, and Professor H. W. Wilson. I owe special debts of gratitude to Mr and Mrs David Michie for their hospitality during my visit to Edinburgh in 1965, and to Mrs Margaret Duthie for suggesting the more specialized study in which this book had its origin.

Acknowledgements are due to the following persons and organizations for permission to reproduce copyright photographs and other illustrative material: Mr Douglas C. Russell (plates 1A,

9A and 9B), Glasgow University Library (plate 1B), the *Glasgow Herald* (plates 2A and 7A), Mr R. M. Pepper (plates 3B, 14A and 16B), the Waterways Museum (plate 4A), Messrs Aerofilms Ltd (plates 4B, 5B, 7B, 16A, 18, 19A and 20A), Falkirk Public Library (plates 5A, 8A and 8B), the Mitchell Library (plate 6A), the *Scotsman* (plate 6B), Mr Eric R. Hatcher (plate 10A), Edinburgh Public Library (plates 10B and 11B), the Paisley Burgh Museum (plates 12A and 12B), Aberdeen Public Library (plates 13A and 13B), the British Museum Newspaper Library (plate 14B), Mr W. A. Sharp (plate 15A), Mr Campbell Gardiner (plate 15B), Mr Tom Weir (plate 17A), Messrs Valentine & Sons (plate 17B), the National Library of Wales (plate 20B). For material included in the maps I have to thank the copyright section of the Ordnance Survey.

My greatest debt is to my husband, who has declined the responsibilities of co-authorship but accepted its labours in full measure.

JEAN LINDSAY

Bangor, 1967

NOTES

Notes to Chapter I

1. Letters of W. Adam, A. Gordon and W. Wishart in the Scottish Record Office; D. Defoe, *A Tour through the Whole Island of Great Britain* (1769), IV, p. 137; J. Knox, *A View of the British Empire* (1785), II, p. 401.
2. *Scots Magazine*, XXIX, p. 129; Knox, op. cit. II, p. 401; J. Smeaton, *Report concerning the Practicability of joining the Rivers Forth and Clyde by a Navigable Canal* (1764).
3. *Journals of the House of Commons*, 16, 20 and 26 March 1767; *Old Statistical Account*, V, p. 586; R. Mackell and J. Watt, *An Account of the Navigable Canal from the Clyde to the Carron* (1767).
4. *Journals of the House of Commons*, 28 April 1767; *Scots Magazine*, XXIX, pp. 131 and 183–4; Mackell and Watt, op. cit.; R. H. Campbell, *Carron Company* (1961), pp. 117–18.
5. Forth & Clyde Canal Minute Book (hereafter FCCMB), 27 May 1767; *Edinburgh Evening Courant*, 6 April 1767; *Scots Magazine*, XXIX, pp. 249–53.
6. J. Smeaton, *Second Report Touching the Practicability of Making a Navigable Canal from the River Forth to the River Clyde* (1767).
7. FCCMB, 22 February 1768; *Journals of the House of Commons*, 11 and 15 December 1767 and 14 January 1768; *Scots Magazine*, XXX, pp. 40 and 386–8.
8. 8 Geo. III, c. 63, 8 March 1768.
9. FCCMB, 7 and 14 March 1768; 8 Geo. III, c. 63; *Edinburgh Evening Courant*, 18 June 1768.
10. Letter of W. Pulteney in the Scottish Record Office; J. Brindley, T. Yeoman and J. Golborne, *Reports Relative to a Navigable Communication Betwixt the Firths of Forth and Clyde* (1768); J. Smeaton, *Review of Several Matters Relative to the Forth and Clyde Navigation* (1768).
11. *Edinburgh Evening Courant*, 2 and 9 November 1768; W. Nimmo, *History of Stirlingshire* (1880), I, p. 283.
12. FCCMB, 7 September 1768; Representation of the Carron Company to the Forth & Clyde Navigation, in the Scottish Record Office.
13. FCCMB, 2 January, 6 February and 7 November 1769 and 29 January 1770; *Scots Magazine*, XXXI, p. 501.
14. FCCMB, 29 January, 1 May, 11 June and 20 July 1770 and 3 and 27 July 1771; 11 Geo. III, c. 62, 8 March 1771; *Scots Magazine*, XXXII, p. 727.
15. FCCMB, 13 September 1771.
16. FCCMB, 5 November and 9 December 1771.
17. FCCMB, 21 December 1771, 14 August 1772 and 27 April and 3 and 4 August 1773.
18. FCCMB, 4 May, 3 and 4 August and 6 December 1773; *Scots Magazine*, XXXV, p. 557.
19. FCCMB, 2 and 4 July, 1 August and 21 September 1775; *Scots Magazine*, XXXVII, p. 54; *Old Statistical Account*, V, p. 587.
20. FCCMB, 5 January, 9 May, 3 October, 26 November and 10 December 1776.
21. FCCMB, 11 March 1778, 11 November 1779 and 21 March 1780; *Glasgow Mercury*, 5 February 1778; *Old Statistical Account*, V, p. 587; J. Priestley, *An Historical Account of the Navigable Rivers, Canals and Railways of Great Britain* (1831), p. 270; H. Hamilton, *Economic History of Scotland in the Eighteenth Century* (1963), p. 237.

22. FCCMB, 6 March 1781 and 29 January and 1 May 1783; *Scots Magazine*, XLV, p. 327; Knox, op. cit. II, pp. 409–10.
23. FCCMB, 24 March and 26 May 1784; 24 Geo. III, c. 59, 19 August 1784.
24. FCCMB, 3 August and 3 December 1784 and 1 February, 8 March, 11 and 20 April and 3 May 1785.
25. FCCMB, 7 June and 27 July 1785 and 25 January 1786; R. Whitworth, *Report Relative to the Tract of the Intended Canal from Stockingfield Westward* (1785), *Survey of the Canal from Glasgow to Grangemouth* (1785) and *Survey of Dolater Bog* (n.d.); A. Millar, *First Report Respecting bringing Supplies of Water through the Monkland Canal* (1785).
26. *Considerations on the State of the Forth & Clyde Navigation* (1786).
27. Letter of Patrick Colquhoun in the Scottish Record Office; FCCMB, 4 October 1786, 8 June 1787, 6 February 1788 and 4 March and 1 April 1789; 27 Geo. III, c. 20, 21 May 1787 and c. 55, 28 May 1787; *Scots Magazine*, LI, p. 516; J. Hopkirk, *Account of the Forth & Clyde Canal* (1816); J. Maxwell, *The Great Canal* (1788); R. Whitworth, *Report Relative to Supplies of Water* (1786).
28. FCCMB, 20 May 1789; *Glasgow Mercury*, 15 September 1789; *Scots Magazine*, LI, p. 516; 30 Geo. III, c. 73, 9 June 1790.
29. FCCMB, 11 November 1789 and 3 February and 18 June 1790; *Glasgow Mercury*, 15 December 1789; B. Woodcroft, *A Sketch of the Origin and Progress of Steam Navigation* (1848), p. 37.
30. FCCMB, 27 February 1782 and 1 October 1788; *Scots Magazine*, LII, p. 409; A. Whyte and D. MacFarlan, *General View of the County of Dunbarton* (1811), p. 265; *Inventory of the Ancient Monuments of Stirlingshire* (1963), II, p. 437.
31. FCCMB, 4 August 1790, 1 June 1791 and 29 November 1792; *Glasgow Courier*, 18 October 1791 and 3 May 1792; *Glasgow Mercury*, 11 September 1792; Hopkirk, op. cit. p. 75; J. Cleland, *Annals of Glasgow* (1816), I, p. 301.
32. FCCMB, 10 May 1792, 2 May and 7 November 1793, 9 January, 26 March and 11 December 1794 and 14 September 1795; *Glasgow Courier*, 28 March 1796; Hopkirk, op. cit. p. 75.
33. FCCMB, 11 February, 14 July and 24 November 1796, 7 May, 12 July and 20 December 1798, 9 and 10 April 1800 and 12 and 18 March 1801; 39 Geo. III, c. 71, 12 July 1799.
34. FCCMB, 5 June 1800 and 26 February 1808; *Glasgow Courier*, 31 March 1803; W. Muir, *Poems on Various Subjects* (1818), p. 10; Woodcroft, op. cit. p. 53.
35. FCCMB, 5 January 1799, 19 January and 12 April 1804, 14 August 1806, 13 February and 17 June 1807, 26 February 1808, 10 May 1809, 13 April 1810 and 12 and 21 February, 25 May, 13 August and 12 November 1812; 46 Geo. III, c. 120, 12 July 1806; 49 Geo. III, c. 74, 20 May 1809; 54 Geo. III, c. 195, 14 July 1814; *Ancient Monuments of Stirlingshire*, II, p. 437; Muir, op. cit. p. 296.
36. FCCMB, 10 October 1814 and 26 April 1815; A. Laird, *General View of the Agriculture of Stirlingshire* (1812), pp. 352–6.
37. FCCMB, 20 March 1816; *Memorial for Kirkman Finlay and Others against the Forth & Clyde Navigation* (1815); *Memorial for the Forth & Clyde Navigation against Kirkman Finlay and Others* (1815); *Observations by the Governor and Council of the Forth & Clyde Navigation* (1816); *Reply to the Observations by the Governor and Council* (n.d.); Hopkirk, op. cit p. 75.
38. FCCMB, 8 August 1799, 18 January 1802 and 26 February 1808; *Glasgow Courier*, 16 June 1796 and 2 May 1818; Hopkirk, op. cit. pp. 47 and 75; Scrapbook in Airdrie Public Library.
39. FCCMB, 4 and 12 March 1818, 17 November 1819, 9 March and 3 April 1820 and 18 May and 14 June 1821; 1 Geo. IV, c. 48, 8 July 1820; *Edinburgh Philosophical Journal*, II, p. 222.
40. FCCMB, 14 February and 8 August 1822, 30 December 1823 and 12 February 1824.
41. FCCMB, 11 March and 9 September 1824, 16 March and 29 April 1825; 3 February 1826 and 14 September and 5 November 1827; 6 Geo. IV, c. 117, 10 June 1825; *Royal Commission on Canals and Waterways* (1906–9), III, p. 142.
42. FCCMB, 27 February, 9 July and 9 October 1828, 6 January, 26 April, 24 June,

14 July and 9 November 1830 and 1 March 1831; W. Fairbairn, *Remarks on Canal Navigation* (1831), pp. 6–7, 14–16, 19 and 24–36; C. L. D. Duckworth and G. E. Langmuir, *Clyde River and Other Steamers* (1946), pp. 174–5.

43. British Transport Historical Records (hereafter BTHR), FCN, 4/1, p. 14; FCCMB, 1 March and 27–8 April 1831, 9 February and 14 June 1832, 20 April and 25 September 1833, 8 April 1835 and 4 February 1836; T. Grahame, *Essays on Inland Communication* (1835), pp. 6, 14–19, 55 and 59–60; J. C. Apperley, *Nimrod's Northern Tours* (n.d.), pp. 279–80.
44. BTHR, FCN, 4/1, p. 14; FCCMB, 27 October 1836, 13 September 1837 and 5 September 1838; 26 Geo. II, c. 96, 7 June 1753; 27 Geo. III, c. 56, 21 May 1787; 6 & 7 Will. IV, c. 51, 20 May 1836; 56 & 57 Vic., c. 179, 27 July 1893; *New Statistical Account* VIII, p. 24.
45. FCCMB, 28 August 1839, 6 November 1840 and 13 May 1842; 4 & 5 Vic., c. 55, 21 June 1841; 5 Vic., c. 41, 18 June 1842; information supplied by Mr D. McKay.
46. FCCMB, 20 April 1833, 13 and 29 October 1842, 24 February, 11 April and 12 July 1843, 19 January, 17 April and 23 September 1844, 18 February 1845 and 20 February, 3 March and 28 October 1846; 8 & 9 Vic., c. 3, 8 May 1845 and c. 148, 21 July 1845; 9 & 10 Vic., c. 147, 3 July 1846; *Petition of the Merchants and Traders in the City of Glasgow* (1843); *Edinburgh and Glasgow Railway Directors' Report* (1846).
47. FCCMB, 27 April 1847 and 11 February, 14 May and 24 October 1849; 9 & 10 Vic., c. 384, 18 August 1846; 12 & 13 Vic., c. 39, 26 June 1849.
48. FCCMB, 24 April 1850, 20 January and 5 May 1851, 14 September and 27 October 1852 and 15 August 1854; 15 Vic., c. 45, 28 May 1852.
49. FCCMB, 12 December 1854, 25 September 1855, 23 April 1856, 28 April 1857, 26 January, 28 April, 6 July and 2 August 1858 and 27 April and 26 October 1859.
50. FCCMB, 23 November 1859, 25 April 1860 and 23 October 1861; 29 & 30 Vic., c. 256, 23 July 1866; 30 & 31 Vic., c. 106, 20 June 1867; Reports of the Governor and Council of the Forth & Clyde Navigation 1864–7; *Proceedings of the Institution of Civil Engineers*, XXVI, pp. 10–14.
51. *Royal Commission on Canals and Waterways*, I ii, p. 11 and III, pp. 437–45.
52. *Royal Commission on Canals and Waterways*, III, pp. 141–2, 147–8 and 441–3 and VII, p. 45.
53. *Weekly Scotsman*, 21 October 1911; 1939 advertising leaflet in my possession; *History of the Forth & Clyde, Monkland and Forth & Cart Canals* (1926); E. A. Pratt, *Scottish Canals and Waterways* (1922), p. 126.
54. BTHR, FCN, 4/1, p. 30; *Glasgow Herald*, 12 January 1920 and 4 July 1923; *Report of the Forth & Clyde Ship Canal Group* (1946); *Clyde Valley Regional Plan* (1949), p. 237; Duckworth and Langmuir, *Clyde River and Other Steamers*, p. 175.
55. Hansard, 2 July 1957; *Glasgow Herald*, 8 June 1951; *Canals and Waterways Report of the Board of Survey* (1955), pp. 184–5.
56. *Report of the Committee of Inquiry into Inland Waterways* (1958), pp. 196–205 and 216–18.
57. 10 & 11 Eliz. II, c. 16, 29 March 1962; Hansard, 30 January 1962; *Glasgow Herald*, 25 February 1959.

Notes to Chapter II

1. J. Watt, *A Scheme for Making a Navigable Canal from the City of Glasgow to the Monkland Collieries* (n.d.); J. P. Muirhead, *Life of James Watt* (1859), pp. 197–8; Hamilton, *Eighteenth Century*, p. 209.
2. G. Thomson, 'James Watt and the Monkland Canal', *Scottish Historical Review*, XXIX, No. 108; J. Coutts, *A History of the University of Glasgow* (1909), p. 266.
3. 10 Geo. III, c. 105, 12 April 1770.
4. Muirhead, op. cit. p. 199; G. Thomson, op. cit.
5. Muirhead, op. cit. p. 199; G. Thomson, op. cit.

6. G. Thomson, op. cit.
7. Muirhead, op. cit. p. 200; 53 Geo. III, c. 75, 21 May 1813; A. Miller, *The Rise and Progress of Coatbridge* (1864), p. 4.
8. *Glasgow Mercury*, 11 May and 29 June 1780 and 7 June 1781; *Scots Magazine*, XLVI, p. 388; *Papers Written in Opposition to the Union Canal* (1817), p. 109; Hamilton, *Eighteenth Century*, p. 239.
9. 30 Geo. III, c. 73, 1790; R. Whitworth, *First Report respecting Bringing Supplies of Water through the Monkland Canal* (1785); Maxwell, op. cit.
10. FCCMB, 7 November 1793; *Scots Magazine*, LVI, p. 449; Hamilton, *Eighteenth Century*, p. 210.
11. *Scots Magazine*, LIII, p. 304 and LVII, p. 747.
12. FCCMB, 5 April and 11 September 1794, 14 July 1796, 9 February 1797 and 10 May, 6 October and 20 December 1798.
13. FCCMB, 19 January 1804 and 14 August and 18 September 1806; 53 Geo. III, c. 73, 21 May 1813; *Papers in Opposition to the Union Canal*, pp. 7, 107 and 109; *Duncan's Itinerary of Scotland* (1827), Appendix 3; *Royal Commission on Canals and Waterways*, IV, p. 423; J. Naismith, *General View of the Agriculture of Clydesdale* (1798), p. 142; J. Headrick, *General View of the Agriculture of the County of Angus* (1813), p. 532; D. Brewster, *The Edinburgh Encyclopaedia* (1830), XV, p. 264; F. H. Groome, *Ordnance Gazetteer of Scotland* (1901), p. 1173; Miller, op.cit. p. 6.
14. FCCMB, 30 December 1823, 6 November 1828 and 6 January 1830; 5 Geo. IV, c. 49, 17 May 1824; *New Statistical Account*, VI, p. 664; H. Hamilton, *The Industrial Revolution in Scotland* (1966), p. 246.
15. FCCMB, 29 September 1832, 25 September 1833 and 24 September 1834; *Coatbridge Advertiser*, 2 February 1861; *New Statistical Account*, VI, pp. 205–6; Ordnance Survey 1st Edition (Lanarkshire) Sheet 7 & 8; *Duncan's Itinerary of Scotland*, p. 111 and Appendix 3; Groome, op. cit. p. 1173; Fairbairn, *Canal Navigation*, p. 61; Cleland, op. cit. I, pp. 310–13.
16. FCCMB, 20 October 1835, 2 November 1837, 3 April 1839, 13 April and 1 October 1841, 10 January 1843 and 20 February and 3 March 1846; 4 & 5 Vic., c. 54, 21 June 1841; 5 Vic., c. 41, 18 June 1842; 6 & 7 Vic., c. 63, 4 July 1843; *Remarks on the Opposition of Sir John Marjoribanks to the Union Canal Bill* (1815); *J. Leslie, Description of an Inclined Plane for Conveying Boats at Blackhill* (1852).
17. BTHR, FCN, 4/1, pp. 27–8; FCCMB, 20 February, 3 March and 20 August 1846 and 26 February 1849; 6 & 7 Vic., c. 63, 4 July 1843; 9 Vic., c. 11, 14 May 1846; 9 & 10 Vic., c. 147, 3 July 1846; *New Statistical Account*, VI, p. 664; *Royal Commission on Canals and Waterways*, III, p. 143.
18. BTHR, FCN, 4/1, p. 10; FCCMB, 20 August 1846, 27 October 1847, 25 October and 20 November 1848, 26 February, 14 May, 3 September and 8 and 17 October 1849 and 13 August 1850; Union Canal Minute Book (hereafter UCMB), 10 August 1847; J. Leslie, op. cit. pp. 5–6; W. M. Acworth, *The Railways of Scotland* (1890), 12–13; Groome, op. cit. p. 1174; J. Cowan, *Glasgow's Treasure Chest* (1951), p. 38.
19. FCCMB, 7 September and 8 October 1850, 22 October 1851, 28 April 1852, 25 January, 28 April and 17 October 1854 and 25 September 1855; Reports of the Governor and Council of the Forth & Clyde Navigation 1859–62; Hamilton, *Industrial Revolution*, 220–1.
20. BTHR, FCN, 4/1, pp. 3, 10 and 20; 30 & 31 Vic., c. 106, 20 June 1867; *Glasgow Herald*, 13 March 1943 and 21 March 1944; Reports of the Governor and Council of the Forth & Clyde Navigation 1866–7; *Royal Commission on Canals and Waterways*, III, pp. 141, 145, 437, 439 and 440; *Proceedings of the Institution of Civil Engineers*, XXVI, p. 12; Miller, op. cit. pp. 12–13, 20–1 and 61–2.
21. 51 & 52 Vic., c. 25, 10 August 1888; *Glasgow Herald*, 21 August 1942, 8 March 1943, 21 March 1944, 23 June 1948 and 12 June and 2 July 1952; *Clyde Valley Regional Plan*, p. 238.
22. *Glasgow Herald*, 2 November and 3 December 1953, 9 July and 6 August 1954, 6 July 1960, 4 August 1961 and 12 and 18 July 1963.

Notes to Chapter III

1. *Glasgow Courier*, 17 January 1793; *Scots Magazine*, LIII, p. 99, LIV, p. 150 and LV, p. 44; J. Grieve and J. Taylor, *Report on the Canal between Edinburgh and Glasgow* (1794); J. Rennie, *Report Concerning the Different Lines Surveyed by John Ainslie and Robert Whitworth Junior* (1797); *Observations on the Report of Mr. Hugh Baird* (1814); Hamilton, *Eighteenth Century*, pp. 207 and 240.
2. Rennie, *Report on Different Lines; Observations on the Report of Mr. Hugh Baird.*
3. H. Baird, *Report on the Proposed Edinburgh & Glasgow Union Canal* (1813).
4. *Observations on the Report of Mr. Hugh Baird*; Hamilton, *Industrial Revolution*, p. 198.
5. J. Grahame, *Answer to a Pamphlet entitled 'Observations on the Report of Mr. Hugh Baird'* (1814).
6. UCMB, 20 April and 17 August 1814; *Scots Magazine*, LXXVI, p. 807; R. Bald, *Report of a Mineral Survey along the Track of the Proposed North or Level Line* (1814).
7. UCMB, 4 and 17 November and 13 and 24 December 1814; H. Baird, *Subsidiary Report to the General Committee* (1814); R. Stevenson, *Report Relative to the Canal between Edinburgh and Glasgow* (1814).
8. UCMB, 17 August 1814; *Scots Magazine*, LXXVI, p. 686 and LXXVIII, pp. 252 and 326; *Remarks on the Opposition of Sir John Marjoribanks to the Union Canal Bill.*
9. *Scots Magazine*, LXXVII, pp. 242–52 and 337.
10. UCMB, 17 June 1815, 3 October and 12 December 1816 and 8 March 1817; *Papers in Opposition to the Union Canal; Observations by the Union Canal Committee* (n.d.).
11. UCMB, 6 June 1817; 57 Geo. III, c. 56, 27 June 1817.
12. UCMB, 5, 6 and 29 August, 4 September, 3 and 7 October and 18 November 1817 and 27 January and 31 March 1818; W. Roughhead, *Burke and Hare* (1948), p. 12.
13. UCMB, 9 November and 15 and 29 December 1818 and 4 September 1821; *Ceremony at the Commencement of the Union Canal* (1818).
14. UCMB, 21 May 1818, 7 March and 3 October 1820, 6 March 1821, 26 January 1822 and 16 April 1830; 59 Geo. III, c. 29, 19 May 1819; *The Tourist's Manual* (1835), p. 39.
15. UCMB, 3 April and 15 December 1818, 1 June 1819, 7 March 1820 and 6 March and 10 August 1821; 1 & 2 Geo. IV, c. 122, 23 June 1821.
16. UCMB, 3 and 31 October and 20 December 1820, 6 and 27 March, 2 October and 18 December 1821 and 2 January, and 5 and 28 March 1822.
17. *Edinburgh Evening Courant*, 24 January and 9 May 1822.
18. UCMB, 1 October 1822 and 4 March and 24 June 1823; 1 & 2 Geo. IV, c. 122, 23 June 1821 and 4 Geo. IV, c. 18, 12 May 1823; *Inventory of the Ancient Monuments of Stirlingshire*, II, p. 438.
19. UCMB, 5 March, 30 April, 3 October and 9 December 1822 and 17 and 31 March, 28 July and 5 September 1823; *Edinburgh Evening Courant*, 9 May 1822.
20. UCMB, 30 April and 31 August 1822 and 28 July, 18 August and 22 September 1823; *A Companion for Canal Passengers Betwixt Edinburgh and Glasgow* (1823).
21. UCMB, 7 April 1823, 29 March and 12 April 1824, 8 October 1827 and 3 April 1829; *Scots Magazine*, XIV, p. 119.
22. UCMB, 31 January and 30 May 1826, 24 September 1827, 7 September 1832 and 4 January and 2 July 1833; 7 Geo. IV, c. 45, 5 May 1826; Hamilton, *Industrial Revolution*, p. 278.
23. UCMB, 16 October and 15 December 1829, 8 January, 10 September and 12 October 1830, 29 April 1831, 3 February, 7 August and 7 September 1832, 21 and 22 March and 2 July 1833 and 3 June 1834.
24. UCMB, 12 June 1834, 27 January, 4 August and 29 September 1835, 1 March, 21 June and 27 December 1836, 7 August 1838 and 8 January 1839.

25. UCMB, 8 January 1830, 20 February and 12 December 1835, 4 December 1838, 21 July, 30 October and 18 December 1840, 10 November 1841, 18 January 1842 and 31 February 1843; 7 & 8 Vic., c. 70, 4 July 1844; *Inventory of the Ancient Monuments of Stirlingshire*, II, p. 439.
26. UCMB, 25 March 1831, 18 May 1832, 7 March, 25 July and 31 October 1837, 7 March, 4 April, 4 August and 20 December 1843, 13 April, 28 May, 14 June, 16 July and 7 October 1844, 9 May 1845 and 3 March and 22 April 1848; Hamilton, *Industrial Revolution*, p. 247.
27. UCMB, 7 and 8 July and 13 November 1846 and 27 July 1847; FCCMB, 14 April 1847; *Edinburgh & Glasgow Railway Directors' Report* (1846).
28. UCMB, 7 April 1849; 12 & 13 Vic., c. 39, 26 June 1849.
29. *Edinburgh Evening Dispatch*, 19 October 1921; *Edinburgh Evening News*, 17 September 1937; *Scotsman*, 15 October 1912; *Weekly Scotsman*, 26 October 1912; *Scottish Geographical Magazine*, 16 April 1923; Hansard, 25 January 1955; *Royal Commission on Canals and Waterways*, III, pp. 150, 437 and 445–6 and VIII, p. 46; *North British Railway Directors' Reports*, 31 July 1870, 31 January 1871, 31 July 1880, 31 January 1881, 31 July 1890, 31 January 1891, 31 July 1900 and 31 January 1901.
30. *Edinburgh Evening Dispatch*, 7 March 1951 and 27 January 1955; Hansard, 25 January 1955; *Report of the Board of Survey*, p. 61; *Report of the Committee of Inquiry*, p. 216; information supplied by British Waterways.

Notes to Chapter IV

1. *Glasgow Courier*, 12 February 1803 and 25 October 1804; *Scots Magazine*, LXVIII, pp. 323 and 326; J. Rennie, *Report on a Survey and Plan by John Ainslie* (1804); J. Wilson, *General View of the Agriculture of Renfrewshire* (1812), p. 186; J. Rickman (ed.), *The Life of Thomas Telford* (1838), p. 67; C. E. Stretton, *The Ardrossan Canal and Railway* (n.d.).
2. *Glasgow Courier*, 24 August 1805; 46 Geo. III, c. 75, 20 June 1806; T. Telford, *Report on the Proposed Canal from Glasgow to the West Coast of the County of Ayr* (1805).
3. Paisley Canal Minute Book (hereafter PCMB), 17 July and 6 November 1806 and 9 April 1807; *Scots Magazine*, LXVIII, p. 323; Cleland, op. cit. I, pp. 314–17; W. Barr, *The Rise and Fall of the Glasgow, Paisley & Ardrossan Canal* (1846); *Correspondence regarding the dismissal of John Giffen* (1866).
4. PCMB, 1 November 1810 and 5 November 1812; *Glasgow Courier*, 8 August 1805; *Exeter Flying Post*, 22 November 1810; *Scots Magazine*, LXXII, p. 873; Wilson, op. cit. p. 188; *Duncan's Itinerary of Scotland*, p. 112; Rickman, op. cit. p. 67; Cleland, op. cit. I, pp. 314–17.
5. PCMB, 25 March 1816, 5 May and 22 July 1817 and 3 February 1820; Letters of William Crossley and others in Glasgow University Library; Letter from Eglinton to Liverpool in the Scottish Record Office; 49 Geo. III, c. 74, 20 May 1809; Cleland, op. cit. I, pp. 314–17.
6. PCMB, 3 June 1816, 12 June and 6 November 1817, 9 August 1819 and 4 November 1824; circular of 1818 in the Baillie Institute, Glasgow.
7. PCMB, 6 November and 5 December 1817, 8 and 15 March 1819, 14 and 28 February 1820, 2 August 1824, 10 May and 23 December 1836 and 17 October 1837; B. Baxter, *Stone Blocks and Iron Rails* (1966), p. 227.
8. PCMB, 6 November 1817, 5 November 1818, 15 January and 26 March 1821, 1 April and 4 November 1822, 6 November 1823 and 28 June and 9 August 1830; Fairbairn, *Canal Navigation*, pp. 18–21.
9. FCCMB, 27 and 28 April 1831; *The Arcana of Science* (1835), p. 68.
10. PCMB, 24 August and 21 September 1830, 28 May 1831, 28 March 1833, 6 February and 15 April 1834, 5 July 1836 and 17 October 1837.
11. FCCMB, 26 November 1834 and 28 August 1840; PCMB, 15 February 1831, 14 August 1832, 17 October and 7 November 1833, 16 May and 18 December 1835

and 28 August 1840; *New Statistical Account*, VII i, pp. 278–9; Rickman, op. cit. p. 69; C. Hadfield, *British Canals* (1950), p. 151; *Burgh of Johnstone Souvenir Brochure* (1951).

12. A. Smith, *Alfred Hagart's Household* (1867).
13. PCMB, 13 February 1826, 2 July 1827, 1 September and 10 November 1828, 7 September 1829, 7 November 1833, 5 November 1835 and 6 June 1836, 7 & 8 Geo. IV, c. 87, 14 June 1827 and 3 & 4 Vic., c. 104, 23 July 1840; *New Statistical Account*, V, pp. 203–4.
14. PCMB, 6 September 1831, 16 September 1834, 6 September 1836, 12 September 1837, 18 September 1838, 22 March 1839 and 17 March 1840; 7 & 8 Geo. IV, c. 87, 14 June 1827; Barr, op. cit.
15. PCMB, 2 March and 3 April 1837; FCCMB, 12 July 1843; *A Bill to authorise the Sale of the Glasgow, Paisley & Johnstone Canal to the Glasgow, Paisley, Kilmarnock & Ayr Railway Company* (1846); *New Statistical Account*, VII, pp. 561–2; Barr, op. cit.; Hamilton, *Industrial Revolution*, p. 248.
16. PCMB, 24 December 1839; *A Bill to authorise the Sale of the Glasgow, Paisley & Johnstone Canal to the Glasgow, Paisley, Kilmarnock & Ayr Railway Company*; Barr, op. cit.
17. Letter of John McInnes in BTHR; 32 & 33 Vic., c. 48, 24 June 1809; Circular issued by McInnes and MacFarlane of Paisley in 1873.
18. Glasgow & South Western Railway Minute Books, 24 January and 9 March 1881; *Minutes of Speech taken before the Select Committee of the House of Commons on the Glasgow & South Western Railway Bill* (1881); 44 & 45 Vic., c. 149, 18 July 1881; G. H. Robin, 'The South Side Suburban Railways of Glasgow' in *The Railway Magazine*, February 1954.

Notes to Chapter V

1. *Aberdeen Journal* (hereafter *AJ*), 28 April and 9 June 1794, 28 July 1795 and 2 February 1796; *Old Statistical Account*, XIX, pp. 226–8.
2. *AJ*, 5 May 1796, 27 February, 6 June and 10 October 1797 and 5 June 1805; *The Aberdeenshire Canal*, a volume of miscellaneous documents in Aberdeen Public Library.
3. *AJ*, 5 June and 15 October 1798, 22 April, 8 June and 7 October 1799 and 24 February and 7 April 1800.
4. *AJ*, 8 December 1800 and 12 January and 2 March 1801; *The Aberdeenshire Canal*.
5. *AJ*, 2 January and 11 August 1802 and 23 February and 9 March 1803.
6. *AJ*, 23 February and 18 May 1803 and 25 January, 1 February and 4 April 1804; *The Aberdeenshire Canal*.
7. *AJ*, 24 April 1805.
8. *AJ*, 5 June 1805; *New Statistical Account*, XII, pp. 68–9; W. Thom, *The History of Aberdeen* (1811), II, p. 177.
9. *AJ*, 11 September 1805, 16 April and 29 October 1806, 8 April 1807 and 20 September 1809; G. S. Keith, *A General View of the Agriculture of Aberdeenshire* (1811), p. 544.
10. *AJ*, 29 July and 2 September 1807 and 27 April, 11 May, 21 September and 7 December 1808; *The Aberdeenshire Canal*.
11. 49 Geo. III, c. 3, 13 March 1809; *The Aberdeenshire Canal*.
12. *AJ*, 19 July and 20 December 1809, 7 December 1814 and 29 January 1817.
13. *AJ*, 27 September 1809, 22 March 1810, 20 and 27 July 1814 and 9 October 1816; *New Statistical Account*, XII, p. 69; R. Southey, *Journal of a Tour in Scotland in 1819* (1929), pp. 78–9.
14. *AJ*, 9 December 1818, 14 April 1819 and 18 February 1829.
15. *AJ*, 2 February 1820, 13 April 1825 and 18 August 1830; *New Statistical Account*, XII, pp. 660 and 683; *The Aberdeenshire Canal*.
16. *AJ*, 9 October 1833 and 28 May and 4 June 1834; *New Statistical Account*, XII, p. 69.

17. Sederunt Book No I of the Great North of Scotland Railway, 3 July, 14 August, 2 September and 14 October 1845; *AJ*, 20 March 1844; *New Statistical Account*, XII, pp. 69–70; *The Aberdeenshire Canal.*
18. Sederunt Book No I of the Great North of Scotland Railway, 2 September and 14 October 1845, 21 March and 27 August 1846 and 29 December 1848.
19. *AJ*, 5 December 1849 and 13 October and 1 December 1852.
20. Sederunt Book No I of the Great North of Scotland Railway, 7, 9 and 10 November 1853; J. Milne, *Aberdeen* (1911), pp. 346–7.
21. Sederunt Book No I of the Great North of Scotland Railway, 7, 10 and 11 November and 2 December 1853 and 16 January, 17 February and 3 March 1854; *The Aberdeenshire Canal.*
22. *AJ*, 6 and 20 September 1854; Keith, op. cit. p. 542; H. A. Vallance, *The Great North of Scotland Railway* (1965), p. 23.

Notes to Chapter VI

1. Letter of J. Watt, Memorial of the Magistrates of Glasgow, Report by J. Watt and Report by G. Clerk Maxwell, all in the Scottish Record Office.
2. Report by G. Clerk Maxwell; *Third Report on the State of the British Fisheries* (1785), X, pp. 126 and 196; Knox, op. cit. pp. 413 and 417.
3. Report by R. Frazer to the British Fishery Society, in the Scottish Record Office; *Prospectus of the Advantages to be Derived from the Crinan Canal* (1792), pp. 13–14; *Further Report on the British Herring Fisheries* (1798), p. 261.
4. *Prospectus of the Crinan Canal*, pp. 1–8.
5. Crinan Canal Records (hereafter CCR), 17 and 31 January and 7 February 1793; unsigned letter in the Scottish Record Office.
6. Letter of the Duke of Argyll in the Scottish Record Office; CCR, December 1792.
7. CCR, 21 February 1793; 33 Geo. III, c. 104, 8 May 1793.
8. List of Crinan proprietors in BTHR; CCR, 16 and 24 May and 19 December 1793, 7 February, 21 March and 4 and 11 December 1794 and 21 January 1795; A. Gibb, *The Story of Telford* (1935), p. 138.
9. CCR, 24 January, 27 March and 27 May 1794, 28 May and 4 December 1795 and 4 February 1796.
10. CCR, 20 June 1794 and 31 March 1796; *Further Report on the British Herring Fisheries*, p. 262.
11. *Further Report on the British Herring Fisheries*, pp. 261–2.
12. CCR, 4 April and 10 and 24 May 1799 and 3 April 1800; 39 Geo. III, c. 27, 10 May 1799 and c. 71, 12 July 1799.
13. CCR, 20 March and 20 and 28 June 1800 and 5 March and 28 May 1801; *Report of the Select Committee on the Caledonian and Crinan Canals* (1839), p. 63.
14. CCR, 18 July 1801, 1 April 1802 and 31 March 1803; Pratt, op. cit. p. 62.
15. CCR, 25 February 1797, 30 April, 31 July and 31 December 1799 and March 1806.
16. CCR, 31 December 1793, 31 July and 31 August 1795, 25 February 1797, 31 May 1798 and 31 January 1812; *Royal Commission on Canals and Waterways*, II, p. 73.
17. CCR, 31 December 1799 and September 1803; A. Murray, *All that is Worth Seeing in Scotland* (1866), p. 35.
18. CCR, 10 April 1804, January and 6 July 1805, 19 May 1806, 16 April 1807, June 1808 and February and 20 August 1809; 45 Geo. III, c. 85, 2 July 1805.
19. CCR, 17 January, 30 and 31 May and 26 June 1811 and 5 February, June and 3 July 1812; 51 Geo. III, c. 117, 26 June 1811.
20. CCR, 29 February and 30 November 1812 and 30 January 1813.
21. CCR, January 1813; *Report of the Select Committee on the Crinan Canal* (1816).
22. CCR, 22 October 1814, April, May and October 1815 and February and May 1816.

23. CCR, January 1813 and 4 and 25 April and 18 June 1816; 56 Geo. III, c. 135, 1 July 1816; Telford, *Report on the Crinan Canal.*
24. *Report of the Select Committee on the Caledonian and Crinan Canals* (1839), p. 63.
25. CCR, December 1816 and January and March 1817.
26. CCR, April, May, August, September and November 1817, December 1818, April and 22 November 1819 and 29 March, 27 September and 15 December 1820; Hamilton, *Industrial Revolution*, p. 215; C. L. D. Duckworth and G. E. Langmuir, *West Highland Steamers* (1967), pp. 3–4.
27. CCR, August 1821, March and 23 May 1822, 14 June and December 1823 and 25 June 1824; Caledonian Canal Minute Book (hereafter CCMB), 31 May 1824; R. L. Mackie, *A Short History of Scotland* (1930), p. 406; A. McQueen, *Echoes of Old Clyde Paddle-Wheels* (1924), p. 28; Duckworth and Langmuir, *West Highland Steamers*, pp. 3–4.
28. CCR, January 1828, May 1829 and 9 August 1835; CCMB, 30 June 1820, 7 June 1825 and 22 February 1826.
29. CCR, 25 March 1834; *Reports of the Caledonian Canal Commissioners* (hereafter RCCC), 1834.
30. CCR, 8 May and 29 August 1835; CCMB, 18 September 1835; RCCC, 1836.
31. CCMB, 23 December 1837 and 18 January 1838; RCCC, 1837 and 1838; J. Walker, *Report on the Caledonian Canal* (1838); Duckworth and Langmuir, *West Highland Steamers*, pp. 5 and 7–8.
32. *Report of the Select Committee on the Caledonian and Crinan Canals* (1839), pp. 57–102.
33. CCR, 17 May 1840 and 6 February 1841; *Report of the Select Committee on the Caledonian and Crinan Canals* (1840), p. 5.
34. CCR, 16 July and 12 November 1846; RCCC, 1841–6; 9 & 10 Vic., c. 362, 7 August 1846 and 12 & 13 Vic., c. 13, 24 May 1849; *Third Statistical Account* (*Argyll*), p. 52.
35. CCMB, 5 June and 24 November 1844 and March 1845; 11 & 12 Vic., c. 54, 14 August 1848.
36. CCR, 17 August 1847; *The Times*, 20 September 1847; A. Helps (ed.), *Leaves from the Journal of our Life in the Highlands* (1868), pp. 47 and 54; Duckworth and Langmuir, *West Highland Steamers*, p. 17.
37. CCR, 4 February 1851; RCCC, 1848–51.
38. CCMB, 18 July 1854; RCCC, 1852, 1855 and 1857; 20 & 21 Vic., c. 27, 10 August 1857; McQueen, op. cit. pp. 55–7; *One Hundred Years: McBrayne's Centenary* (1951); Duckworth and Langmuir, *West Highland Steamers*, p. 30.
39. CCMB, 1 August 1858; RCCC, 1858.
40. CCMB, 11 and 21 March and 6 December 1859 and 4 January and 23 February 1860; RCCC, 1859 and 1864; 23 & 24 Vic., c. 46, 23 July 1860.
41. RCCC, 1863, 1866 and 1869; Duckworth and Langmuir, *West Highland Steamers*, pp. 36–9.
42. Minutes of Evidence for the Inquiry into Crinan Canal Dues, in the Scottish Record Office; CCMB, 5 March 1877; RCCC, 1879–81; K. Baedeker, *Great Britain* (1901), p. 533.
43. CCMB, 4 December 1884; RCCC, 1881 and 1883–4; 46 & 47 Vic., c. 123, 16 July 1883.
44. CCMB, 30 April 1868, 4 August 1885, 20 August 1887, 24 July 1889, 25 July 1890 and 4 August 1905; RCCC, 1890 and 1892; 50 & 51 Vic., c. 173, 8 August 1887 and 55 Vic., c. 12, 20 May 1892.
45. *Royal Commission on Canals and Waterways*, I ii, pp. 71–6.
46. *Royal Commission on Canals and Waterways*, I ii, pp. 71–6.
47. *Royal Commission on Canals and Waterways*, V ii, pp. 1–10.
48. *Royal Commission on Canals and Waterways*, III, pp. 381, 395, 399–401 and 410–11.
49. Letter from the Convener of the Argyll County Council in the Scottish Record Office; *Glasgow Herald*, 13 August 1907; *Scottish Bankers' Magazine*, 1913; *Royal Commission on Canals and Waterways*, V ii, p. 198 and VII, p. 185.
50. RCCC, 1916–18 and 1920; 8 Geo. V, c. 50, 15 August 1919.

51. *Report of the Crinan Canal Committee* (1921); *Glasgow Herald*, 28 July 1934; Duckworth and Langmuir, *West Highland Steamers*, pp. 55 and 75.
52. *Report of the Board of Survey*, p. 71; *Report of the Committee of Inquiry*, pp. 193–5 and 212–13; *Glasgow Herald*, 2 May 1930, 10 November 1933 and 6 July and 7 August 1934; *One Hundred Years: MacBrayne's Centenary*; Duckworth and Langmuir, *West Highland Steamers*, p. 39.
53. 10 & 11 Eliz. II, c. 46, 1 August 1962; information supplied by British Waterways.

Notes to Chapter VII

1. *Third Report on the State of the British Fisheries* (1774), X, p. 114; *Third Report on the State of the British Fisheries* (1785), p. 42; *Old Statistical Account*, XXI, p. 293; J. Knox, op. cit. p. 433; R. Jamieson (ed.), *Letters from a Gentleman in the North of Scotland* (1818), II, p. 211–12; Muirhead, op. cit. p. 212.
2. MS. Queries about a Loch Shiel Canal, in the Scottish Record Office; T. Telford, *A Survey and Report of the Coasts and Central Highlands of Scotland* (1803); L. T. C. Rolt, *Thomas Telford* (1958), p. 79.
3. 43 Geo. III, c. 102, 27 July 1803.
4. CCMB, 6 August 1803; RCCC, 1804.
5. 44 Geo. III, c. 62, 29 June 1804.
6. RCCC, 1805; Rickman, op. cit. p. 262; Rolt, op. cit, p. 81; C. Hadfield, *The Canals of the West Midlands* (1966), p. 173.
7. Letter of T. Telford in the Scottish Record Office; *Scots Magazine*, LXXI, p. 101.
8. RCCC, 1805.
9. RCCC, 1805.
10. Petition to the Commissioners for the Caledonian Canal, in the Scottish Record Office; RCCC, 1806 and 1807.
11. CCMB, 2 July 1832; RCCC, 1808; Rickman, op. cit. p. 63.
12. RCCC, 1809; Rolt, op. cit. p. 84.
13. RCCC, 1810 and 1811; R. Southey, *Journal of a Tour in Scotland* (1929), p. 205.
14. RCCC, 1811.
15. CCMB, 27 February 1812; RCCC, 1811; Rolt, op. cit, p. 86.
16. RCCC, 1812 and 1813.
17. Letter of William Grant in the Scottish Record Office; CCMB, 10 July 1807, 27 March 1813 and 21 May 1814.
18. RCCC, 1814.
19. RCCC, 1814; J. G. Lockhart, *Memoirs of the Life of Sir Walter Scott* (1839), IV, p. 341.
20. Summons of Declarator by Col A. R. MacDonnell against the Caledonian Canal Commissioners, in the Scottish Record Office; CCMB, 22 June 1816 and 20 February 1817; RCCC, 1815 and 1816; Southey, *Journal of a Tour in Scotland*, p. 187; M. Gray, *The Highland Economy 1750–1850* (1957), pp. 181–8.
21. CCMB, 29 December 1815 and 17 May and 22 June 1816; Gibb, op. cit. p. 222; Rolt, op. cit. pp. 87 and 91.
22. RCCC, 1816; Southey, *Journal of a Tour in Scotland*, 182–3; Rickman, op. cit. p. 61; Gibb, op. cit. p. 222.
23. CCMB, 20 February 1817, 19 March 1818 and 25 May 1820.
24. CCMB, 26 May 1818, 25 May 1819, 25 May 1821 and 2 July 1832.
25. CCMB, 23 May 1822; RCCC, 1821 and 1822; Southey, *Journal of a Tour in Scotland*, p. 189; Brewster, *The Edinburgh Encyclopaedia*, XV, p. 266; Rolt, op. cit. p. 88.
26. RCCC, 1818.
27. Letter of William Young in the Scottish Record Office; Hansard, XXXIX, p. 1119.
28. Miscellaneous papers on the Caledonian Canal in the Scottish Record Office; RCCC, 1819; *Blackwood's Edinburgh Magazine*, VII, pp. 427–36.

29. Report by T. Telford in the Scottish Record Office; RCCC, 1819 and 1821.
30. Letters of James Davidson and Walter Easton in the Scottish Record Office; CCMB, 22 February 1826; *Inverness Courier*, 31 October 1822.
31. Letter of Col. A. R. MacDonnell in the Scottish Record Office; J. Fraser, *Reminiscences of Inverness* (1905); *One Hundred Years: MacBrayne's Centenary.*
32. CCMB, 30 May 1823 and 31 May 1824; RCCC, 1823.
33. CCMB, 10 May and 18 November 1824 and 22 February 1826; RCCC, 1824.
34. CCMB, 18 November 1824; RCCC, 1827 and 1828; 6 Geo. IV, c. 15, 31 March 1825.
35. CCMB, 22 February 1826 and 23 May 1827.
36. CCMB, 30 May 1828.
37. R. Southey, *Poems* (1909), pp. 444–5.
38. Report by G. May in the Scottish Record Office; CCMB, May 1829 and 13 July 1830.
39. CCMB, 29 May 1829, 23 August 1831 and 9 April 1846.
40. CCMB, 21 June and 10 October 1834 and 18 September 1835.
41. RCCC, 1835; *Report of the Select Committee on the Caledonian and Crinan Canals* (1839), pp. 117–20; G. Anderson, *Guide to the Highlands and Islands of Scotland* (1834), p. 265; *The Tourist's Manual* (1835), pp. 219–20.
42. *Report of the Select Committee*, pp. 120–44.
43. *Report of the Select Committee*, pp. v, 105–12 and 144–6.
44. *Report of the Select Committee*, pp. 8–11, 20, 27–38 and 42–5.
45. *Report of the Select Committee*, p. viii.
46. CCMB, 1 September 1841; 3 & 4 Vic., c. 41, 4 August 1840.
47. W. E. Parry, *Report on the Caledonian Canal* (1842).
48. *Further Report of the Select Committee on the Caledonian and Crinan Canals* (1842).
49. Letters of G. May in the Scottish Record Office; CCMB, 2 August 1842; RCCC, 1843 and 1844.
50. Letter of G. May in the Scottish Record Office; CCMB, 20 July 1843; RCCC, 1845.
51. RCCC, 1846 and 1847.
52. CCMB, 8 October 1845; RCCC, 1847; 10 & 11 Vic., c. 208, 9 July 1847.
53. CCMB, 26 August 1848; RCCC, 1847 and 1848; T. Ferguson, *The Dawn of Scottish Social Welfare* (1948), pp. 198–202.
54. RCCC, 1848.
55. CCMB, 16 February, 5 March, 27 April and 1, 14 and 23 May 1849, 19 April 1850 and 18 July 1854.
56. CCMB, 19 April 1850; RCCC, 1850 and 1851; McQueen, op. cit. pp. 55–6.
57. CCMB, 16 July 1850 and 23 June 1852.
58. CCMB, 18 July 1854; RCCC 1852, 1853 and 1854.
59. RCCC, 1857, 1858 and 1860; 20 & 21 Vic., c. 27, 10 August 1857.
60. RCCC, 1860 and 1863; 23 & 24 Vic., c. 46, 23 July 1860.
61. RCCC, 1864, 1866 and 1867; *One Hundred Years: MacBrayne's Centenary*; Duckworth and Langmuir, *West Highland Steamers*, p. 14.
62. CCMB, 21 March 1876; RCCC, 1869, 1870, 1873, 1875 and 1880; J. Fowler, *Report on the Floods of the River Ness* (n.d.).
63. RCCC, 1879, 1880, 1882 and 1883; J. Thomas, *The West Highland Railway* (1965), pp. 38–9.
64. CCMB, 20 November 1888; RCCC, 1868, 1878, 1886 and 1888; *Royal Commission on Canals and Waterways*, I ii, p. 18.
65. RCCC, 1895, 1898 and 1904; Fraser, op. cit. p. 76; A. Mackenzie, *Guide to Inverness* (1893), p. 75; Thomas, op. cit. pp. 78–9.
66. *Royal Commission on Canals and Waterways*, III, pp. 405–6, 426 and 453.
67. *Royal Commission on Canals and Waterways*, I ii, pp. 28 and 77 and VII, p. 184.
68. RCCC, 1911, 1912, 1913, 1915, 1916, 1917, 1918 and 1919; Pratt, op. cit. pp. 35–6.
69. RCCC, 1920; 8 Geo. V, c. 50, 15 August 1919; Pratt, op. cit. pp. 40–2.
70. *Glasgow Herald*, 8 and 14 September 1922, 4 April 1927 23, February 1929, 3

March and 20 April 1931, 2 July 1937 and 26 April and 5 September 1944; *Aberdeen Press and Journal*, 19 September 1966.

71. *Report of the Board of Survey*, pp. 71 and 179–82; *Report of the Committee of Inquiry into Inland Waterways*, pp. 59, 189–91 and 212.
72. 10 & 11 Eliz. II, c. 46, 1 August 1962; *Glasgow Herald*, 31 August 1962; *Modern Transport*, 21 September 1963 and 4 January 1964; T. Weir, 'The Canal that nearly Missed the Boat', *Scottish Field*, November 1966.
73. *Modern Transport*, November 1965; British Waterways Board pamphlet on the Caledonian Canal; information supplied by British Waterways.

Notes to Chapter VIII

1. Undated cutting from the *Dumfries Courier and Herald; New Statistical Account*, IV, p. 397.
2. Dr Singer, *General View of the Agriculture in the County of Dumfries* (1812), pp. 413–14; J. Smeaton, *Reports* (1837), I, pp. 1–3; S. Smiles, *Lives of the Engineers* (1861), II, p. 50; J. Butt, *Industrial Archaeology of Scotland* (1967), p. 175.
3. Minutes of the Glenkens Canal Company, 17 June 1802; 42 Geo. III, c. 114, 26 June 1802; *Old Statistical Account*, IV, pp. 265–6 and VIII, pp. 303–4; G. Chalmers, *Caledonia* (1894), VII, pp. 11–14; information supplied by Mrs M. D. McLean and Mr Jack Howdle.
4. *New Statistical Account*, IV, pp. 292–3 and 375; J. Butt, 'The Industrial Archaeology of Gatehouse-of-Fleet', *Industrial Archaeology*, III, pp. 134–5.
5. W. Aiton, *General View of the Agriculture of the County of Ayr* (1811), p. 561; Hamilton, *Eighteenth Century*, pp. 203–4; information supplied by Mr. John R. Hume.
6. *Glasgow Courier*, 21 September 1797; *New Statistical Account*, V, p. 554; Aiton, op. cit. p. 561; C. F. Dendy Marshall, *A History of British Railways Down to the Year 1830* (1938), pp. 131–3.
7. *Old Statistical Account*, VII, pp. 10 and 15; *New Statistical Account*, V, p. 442; Ordnance Survey 1st Edition (Ayrshire) Sheet 16; Hamilton, *Eighteenth Century*, pp. 186 and 211.
8. Wilson, op. cit. p. 186; Muirhead, op. cit. p. 208.
9. 26 Geo. II, c. 96, 7 June 1753; 27 Geo. III, c. 56, 21 May 1787; 5 & 6 Will. IV, c. 32, 1835; *Scots Magazine*, XLIX, p. 412; *Old Statistical Account*, II, p. 170; W. M. Metcalfe, *A History of Paisley* (1909), pp. 344–6 and 419–21.
10. Report of the Committee of Management of the Forth & Cart Junction Canal Company in BTHR; BTHR, FCN, 4/1, p. 4; FCCMB, 14 February 1799, 25 March 1840 and 24 October 1855; 6 & 7 Will. IV, c. 51, 20 May 1836; 56 & 57 Vic., c. 179, 27 July 1893; *Royal Commission on Canals and Waterways*, I ii, p. 517.
11. FCCMB, 2 February 1837; 1 Vic., c. 111, 12 July 1837.
12. Minutes of the Vale of Leven Canal Company in the Scottish Record Office; FCCMB, 28 January 1842 and 17 April 1844; J. Thomson, *The Proposed Vale of Leven Canal* (1841).
13. Letters of L. Gordon and Sir Neil Menzies in the Scottish Record Office; Notes from Leslie's Report on the Practicability of Rendering the Dochart Navigable for Small Craft, in the Scottish Record Office.
14. *Scots Magazine*, November 1963; information supplied by Mr Jack Howdle.
15. *Old Statistical Account*, X, p. 551; *New Statistical Account*, VII, p. 457; Ordnance Survey 1st Edition (Argyll) Sheets 9, 10 and 11; J. Smith, *General View of the Agriculture of the County of Argyll* (1798), p. 279; Muirhead, op. cit. p. 208; Chalmers, op. cit. VII, p. 11; information supplied by Mr Jack Howdle.
16. Prospectus of the Canal from Loch Awe to the Crinan Canal, in the Scottish Record Office; Smith, op. cit. p. 279.
17. N. Macrae, *The Romance of a Royal Burgh* (1923), pp. 61, 248 and 250; *New Statistical Account*, XIV, pp. 229–30.

18. W. Leslie, *General View of the Agriculture in the Counties of Nairn and Moray* (1811), p. 385; *Scottish Review*, IV, pp. 115–37; Groome, op. cit. p. 372.
19. *Birmingham Locomotive Club Bulletin*, July 1964; *New Statistical Account*, XIII, p. 132.
20. *Old Statistical Account*, IV, p. 57; Chalmers, op. cit., VII, pp. 11–14; Keith, op. cit., p. 10; A. C. O'Dell and K. Walton, *The Highlands and Islands of Scotland* (1962), pp. 201–5; information supplied by Mrs L. Keith and Mr R. F. Bandeen.
21. Letters of the Earl of Findlater and Seafield, J. Watt and G. Young in the Scottish Record Office; Headrick, op. cit. pp. 528–32; Muirhead, op. cit. p. 204.
22. Memorandum of J. Rennie on the Proposed Canal between Arbroath and Forfar, in the Scottish Record Office; G. Robertson, *General View of Kincardineshire* (1810), p. 402; R. Stevenson, *Report relative to the Strathmore Canal between Forfar and Arbroath* (1817); G. Hay, *History of Arbroath* (1876), p. 375.
23. Report of R. Frazer concerning the Funds of the Forfeited Estates, in the Scottish Record Office; Muirhead, op. cit. p. 205; J. Robertson, *General View of the Agriculture of the County of Perth* (1799), p. 428.
24. Letter of R. Frazer, Report of R. Frazer concerning the Funds of the Forfeited Estates, and Circular on the Formation of a Canal from Loch Earn to the Tay, all in the Scottish Record Office; *Scots Magazine*, LXIX, pp. 83–5; Muirhead, op. cit. p. 205.
25. J. Thomson, *General View of the Agriculture of the County of Fife* (1800), pp. 294–5 and 403–6; information supplied by Mrs E. Lindsay.
26. *Old Statistical Account*, IV, p. 543.
27. *Old Statistical Account*, XIV, pp. 621–8; Hamilton, *Eighteenth Century*, p. 107; information supplied by Mr Jack Howdle.
28. Smeaton, *Reports*, I, pp. 96–8, 106–8 and 114.
29. Letter of A. Stirling in the Scottish Record Office; J. Robertson, op. cit. p. 427; Smeaton, *Reports*, I, p. 100.
30. A. McGibbon, *A Report as to Improving the Navigation of the River Forth* (1810), pp. 49, 53, 60 and 97.
31. FCCMB, 4 February 1836; *Prospectus of a Canal intended to connect the Town of Stirling with the Cities of Edinburgh and Glasgow* (1835).
32. *Edinburgh Evening Courant*, 9 November 1782 and 12 March 1783; *Glasgow Courier*, 10 September 1796; 8 Geo. III, c. 63, 8 March 1768; 24 Geo. III, c. 5, 24 December 1783; *New Statistical Account*, II, pp. 138 and 148; Muirhead, op. cit. p. 208; T. J. Salmon, *Borrowstounness and District* (1913), pp. 328–31.
33. R. Somerville, *General View of the Agriculture of East Lothian* (1805), p. 223.
34. Report of the Committee relative to the Proposed Canal between Berwick and Ancrum Bridge, in the Scottish Record Office; *Scots Magazine*, LI, pp. 314 and 614; R. Douglas, *General View of the Agriculture of the Counties of Roxburgh and Selkirk* (1798), p. 205.
35. *Ministry of Transport Report of the Mid-Scotland Ship Canal Committee* (1930); J. Law Crawford, *Forth & Clyde Ship Canal* (1891).
36. *Glasgow Herald*, 28 January and 14 February 1907, 31 December 1908 and 20 January, 29 March and 1 December 1909; *Ministry of Transport Report of the Mid-Scotland Ship Canal Committee*; *Royal Commission on Canals and Waterways*, VII, p. 180.
37. *Glasgow Herald*, 20 October 1911, 21 November 1912, 22 November 1916, 2 June 1917, 11 March and 3 November 1919, 4 January 1921, 28 November 1922, 10 and 30 January and 16 February 1923, 15 February and 7 June 1924, 22 October 1929 and 1 September 1930; *Ministry of Transport Report of the Mid-Scotland Ship Canal Committee*.
38. *Glasgow Herald*, 30 July 1942 and 10 June, 15 and 21 July and 11 and 18 November 1943; *The Ministry of Transport Report of the Forth & Clyde Ship Canal Group* (1946); *Report of the Committee on the Mid-Scotland Ship Canal* (1947).

APPENDIX I

Summary of Facts about the Canals of Scotland

A. *Rivers Successfully Made Navigable*

None; though improvements have been made to such estuaries as those of the Cart, the Peffrey and the Water of Fleet, and vessels have operated on short stretches of the Devon and the Kirkcudbright Dee.

B. *Rivers with Uncompleted Navigation Works*

None

C. *Canals, the Main Lines of which were Completed as Authorized*

Canal	*Date of Act under which Work was Begun*	*Date Opened*	*Approximate Cost at Opening*	*Terminal Points*	*Branches Built*
Aberdeenshire	1796	1805	£43,895	Aberdeen–Inverurie	
Burnturk		*c.* 1800 *c.* 1800		Colliery–Lime-Works Lime-Works–Kingskettle	
Caledonian	1803	1822	£905,258	Corpach–Clachnaharry	
Campbeltown		1794		Argyll Colliery–Campbeltown	
Carlingwark		*c.* 1765 *c.* 1780		Carlingwark Loch–River Dee Old Greenlaw–Glenlochar Bridge	
Crinan	1793	1801	£140,610	Ardrishaig–Crinan	
Edinburgh & Glasgow Union Canal	1817	1822 1823	£461,760	Camelon–Edinburgh	 Port Maxwe
Forth & Cart	1836	1840		River Clyde–Whitecrook	

Length	*Greatest Number of Locks*	*Size of Boats Taken*	*Date of Disuse for Commercial Traffic*	*Date of Abandonment*	*Whether bought by Railway and Present Ownership*
8¼ miles	18	57 ft by 9 ft[1]	1854	1854	Bought in 1845 by the Great North of Scotland Railway
miles	None	4–10 tons burden	*c.* 1830	*c.* 1830	
mile	None	2 tons burden	*c.* 1830	*c.* 1830	
o miles	29	Normally, 160 ft by 36 ft	Used	—	British Waterways Board
miles	None		1856	1856	
½ miles			*c.* 1840	*c.* 1840	
mile			*c.* 1840	*c.* 1840	
miles	15	88 ft by 20 ft	Used	—	British Waterways Board
1½ miles	11	69 ft by 12½ ft	1933	1965	British Waterways Board (and see pp. 83–5)
mile	None	69 ft by 12½ ft	1933	1965	British Waterways Board
mile	3	67 ft by 15 ft	1893	1893	Bought in 1867 by the Caledonian Railway

[1] Size of locks.

Canal	*Date of Act under which Work was Begun*	*Date Opened*	*Approximate Cost at Opening*	*Terminal Points*	*Branches Built*
Forth & Clyde	1768	1790[1]	£330,000	Grangemouth–Bowling	—
		1777	—	—	Stockingfiel Hamiltonh
		c. 1782	—	—	Carron Cut
		c. 1788	—	—	Glassford's
		1791	—	—	Hamiltonhil Port Dund
		1791	—	—	Port Dunda Monkland Basin
Inverarnan	—	—	—	River Falloch–Inverarnan	—
Kilbagie	—	c. 1780	—	Kennetpans–Kilbagie	
Loch Morlich	—	—	—	—	—
Monkland	1770	1793	£120,000	Woodhall–Glasgow	—
		c. 1820	—	—	Calder
		c. 1820	—	—	Dundyvan
		c. 1820	—	—	Gartsherrie
		c. 1820	—	—	Langloan
Muirkirk	—	c. 1789	—	—	—
Sir Andrew Wood's	—	c. 1495	—	Largo House–Largo Church	—
Stevenston	—	1772	—	Stevenston–Saltcoats	—
		1778	—	—	Undergrou Subsidiarie

[1] 1775 to Stockingfield.

Length	*Greatest Number of Locks*	*Size of Boats Taken*	*Date of Disuse for Commercial Traffic*	*Date of Abandonment*	*Whether bought by Railway and Present Ownership*
miles	40	66 ft × $19\frac{2}{3}$ ft	1962	1962	British Waterways Board (and see pp. 47–51)
miles	None	66 ft × $19\frac{2}{3}$ ft	1962	1962	British Waterways Board
nile	None	—	*c.* 1810	*c.* 1810	—
nile	—	—	—	—	—
nile	None	66 ft × $19\frac{2}{3}$ ft	1962	1962	British Waterways Board
nile	1	66 ft × $13\frac{1}{2}$ ft	1962	1962	
nile	None	—	—	—	—
nile	—	—	Before 1861	Before 1861	—
—	—	—	—	*c.* 1850	—
¼ miles	18	66 ft × $13\frac{1}{2}$ ft	*c.* 1935	1950	British Waterways Board (and see pp. 61–5)
nile	None	—	—	—	—
nile	None	—	—	—	—
nile	None	—	—	—	—
nile	None	—	—	—	—
nile	None	—	—	—	—
mile	None	—	Before 1790	Before 1790	—
miles	None	—	*c.* 1830	*c.* 1830	—
			c. 1830	*c.* 1830	—

D. *Canals, the Main Lines of which were not Completed*

Canal	*Date of Act under which Work was Begun*	*Date Opened*	*Approximate Cost of Opening*	*Authorized Terminal Points*	*Terminal Points as Built*
Glasgow Paisley & Johnstone	1806	1811	£130,000	Glasgow–Ardrossan	Glasgow–Johnstone

E. *Canals Partly Built but not Opened*

Canal	*Date of Act under which Work was Begun*	*Authorized Terminal Points*	*Cost of Work Done*	*Date of Abandonmen*
Bo'ness	1768	Grangemouth–Bo'ness	£10,000	1797
Pitfour	*c.* 1790	Scotstoun of St Fergus–Pitfour	—	*c.* 1800

F. *Canals Authorized but not Begun*

Canal	*Date of Act*	*Estimated Cost*	*Authorized Terminal Points*
Glenkens	1802	£33,382	Dalry–Kirkcudbright
Campsie	1837	£21,000	Campsie–Forth & Clyd near Kirkintilloch

Length	*Greatest Number of Locks*	*Size of Boats Taken*	*Date of Disuse for Commercial Traffic*	*Date of Abandonment*	*Whether bought by Railway and Present Ownership*
miles	None	70 ft × 7⅓ ft	1881	1881	Bought in 1869 by the Glasgow & South Western Railway

APPENDIX II

Principal Engineering Works

A. *Inclined Planes*

Canal	*Location of Plane*	*Vertical Rise*	*Dates Working*	*Notes*
Monkland	Blackhill	96 ft	1850–c. 1887	Length of incline 1,040 ft. Tw 7-ft railways. Boats carried i caissons, worked by two stean engines

B. *Outstanding Aqueducts*

Canal	*River Crossed*	*Date of Completion*	*Number of Arches*	*Length*	*Height*
Forth & Clyde	Kelvin	1790	4	400 ft	70 ft
Paisley	Cart (at Blackhall)	1809	1	240 ft	30 ft
Union	Almond	1822	5	420 ft	76 ft
Union	Avon	1822	12	810 ft	86 ft
Union	Water of Leith (at Slateford)	1822	8	500 ft	75 ft

C. *Tunnels over 500 yd*

Canal	*Location of Tunnel*	*Date of Completion*	*Length*
Union	Falkirk	1822	696 yd

APPENDIX III

Traffic Statistics for the Principal Canals

A. *Number of Vessels Passed (Forth & Clyde)*

Year	*Through Passages*		*Return Passages from Grangemouth to Port Dundas*
	West	*East*	
1801	650	636	1,084
1802	619	478	1,259
1803	381	342	1,166
1804	345	368	1,042
1805	390	379	1,237
1806	403	424	1,179
1807	394	423	1,140
1808	324	406	2,860
1817	315	306	1,491
1818	376	394	1,624

B. *Number of Vessels Passed (Caledonian)*

Year	*Through Passages*		*Passages by Steamboats*	*Partial Passages*
	West	*East*		
1830	202	163	221	416
1831	171	152	207	531
1832	182	143	143	778
1833	179	155	162	586
1834	197	131	186	486
1835	160	128	248	494
1836	238	189	235	502
1837	249	216	199	578
1838	245	176	159	576
1839	328	226	175	526

C. *Number of Vessels Passed (Sea-to-Sea Canals)*

Year	Forth & Clyde	Crinan			Caledonian		
	Fishing-boats	Commercial	Fishing	Others	Commercial	Fishing	Others
1951	—	963	712	689	176	569	130
1952	—	945	627	719	146	504	136
1953	189	840	438	697	146	448	144
1954	153	738	462	749	96	546	140
1955	119	787	435	725	118	439	103
1956	98	701	388	701	107	448	141

D. *Number of Passengers Carried*

Year	Monk-land	Forth & Clyde	Paisley	Aberdeen-shire	Union	Caledonian	Crinan
1815	—	85,368	—	—	—	—	—
1828	—	95,836	—	—	—	—	—
1831	25,129	—	79,455	—	—	—	6,571
1832	—	—	148,516	—	—	—	9,594
1833	—	—	240,062	—	—	—	12,777
1834	31,784	—	307,275	—	121,407	—	18,972
1835	—	—	373,290	—	127,292	—	11,344
1836	33,400	197,710	423,186	—	—	—	17,862
1840	—	—	—	c. 10,000	—	—	—
1863	—	—	—	—	—	15,560	—

E. *Tonnage of Principal Cargoes (Forth & Clyde)*

Year	Manure	Timber	Stones	Coal
1800	—	—	—	24,407
1804	—	—	—	47,091
1808	—	—	—	80,647
1823	—	5,080[1]	—	—
1827	—	2,282[1]	—	—
1837	31,506	—	12,799	68,230
1841	26,356	—	28,456	82,410
1846	15,782	—	46,667	63,811
1913	3,206	85,067	8,983	182,324
1925	1,842	53,586	545	31,982

[1] Grangemouth to Port Dundas.

. *Tonnage of Principal Cargoes (Highland Canals)*

Year	Crinan			Caledonian		
	Coal	Liquids	Others	Coal	Liquids	Others
1951	35,264	6,400	24,498	1,083	2,330	25,010
1952	32,423	5,787	26,351	Nil	871	22,727
1953	32,855	4,718	21,258	Nil	308	20,397
1954	33,690	4,551	17,772	Nil	230	11,999
1955	28,703	5,294	17,114	Nil	557	19,175
1956	26,055	5,277	16,893	584	564	24,356

. *Revenue from Passengers and Principal Cargoes (Forth & Clyde), in Pounds*

Year	Passengers	Iron	Grain	Sugar	Coal	Herrings and Salt	Timber
1802	—	1,400	3,471	—	1,801	2,885	4,619
1805	—	2,140	3,841	1,501	3,437	660	4,267
1808	580	2,058	6,022	3,005	4,247	1,181	1,437
1810	2,449	1,582	7,163	6,620	4,472	1,416	4,947
1812	3,453	1,519	4,674	4,186	4,588	1,002	1,986
1814	6,132	1,449	10,903	4,943	3,755	2,043	3,388
1818	7,819	—	13,220	—	4,440	—	4,037
1820	7,810	—	8,287	—	2,680	—	1,417
1822	7,933	—	12,028	—	3,475	—	2,828
1824	5,696	—	12,261	—	3,514	—	5,105
1826	4,217	—	11,466	—	3,643	—	3,048
1828	5,999	—	11,700	—	1,713	—	2,852

I. *Principal Items of Revenue (Crinan), in Pounds*

Year	Slates	Coal	Fishing Trade	General Goods	Dues on Small Boats
1818	241	124	567	371	337
1821	245	125	338	222	254
1824	212	143	308	171	251
1827	262	239	140	214	156
1830	208	197	241	163	173
1843	234	293	112	175	187
1845	292	382	174	311	312
1856	304	399	126	445	51
1858	159	398	177	430	32
1872	616	845	274	683	13

APPENDIX IV

Summary of the Financial History of the Principal Canals

A. *Tonnage (Lowland Canals)*

Year	*Paisley*	*Monkland*	*Forth & Clyde*	*Union*	*Aberdeenshire*
1800	—	—	129,480	—	—
1808	—	—	178,282	—	—
1817	22,865	—	—	—	—
1831	48,191	222,474	—	—	—
1832	51,198	—	307,388	161,939	7,195
1838	—	—	460,396	—	13,563
1846	—	831,600	537,032[1]	—	—
1850	—	1,058,310	1,147,730	—	—
1863	—	1,529,918	—	—	—
1866	93,235	—	2,925,453	—	—
1886	—	—	1,413,277	—	—
1888	—	—	—	129,000	—
1907	—	—	—	117,735	—
1913	—	77,489	697,220	65,118	—
1921	—	30,128	—	19,633	—
1945	—	—	25,000	—	—

[1] April to September.

B. *Annual Revenue and Expenditure (Sea-to-Sea Canals), in Pounds*

Year	*Forth & Clyde*		*Crinan*		*Caledonian*	
	Revenue	*Expenditure*	*Revenue*	*Expenditure*	*Revenue*	*Expenditure*
1802	23,371	9,422	—	—	—	—
1806	26,955	11,877	—	—	—	—
1808	28,214	—	1,155	—	—	—
1810	40,454	—	1,730	—	—	—
1818	42,265[1]	—	1,663	1,132	—	—
1819	41,240	—	2,031	1,503	—	—
1820	37,215	16,178	1,678	1,339	—	—
1821	40,408	13,762	1,656	1,385	—	—
1822	43,647	11,471	1,371	1,466	—	—
1823	44,449	13,525	1,450	1,513	—	—
1824	49,712	13,630	1,766	1,704	2,159	—
1825	45,936[1]	13,169	2,142	1,940	—	—
1826	44,733	13,944	2,007	1,664	—	—
1827	51,868	15,902	1,924	1,591	2,445	3,100
1828	50,123	13,127	1,716	1,470	2,500	—
1829	50,662	14,046	1,552	2,267	2,194	4,834
1830	53,881	16,334	1,806	1,940	2,575	4,573
1831	51,407	18,226	1,511	1,633	2,269	—
1832	40,562	14,899	1,766	1,537	—	—
1833	46,292	13,734	2,049	1,895	—	—
1834	51,167	13,181	1,900	1,412	—	—
1835	—	—	1,704	1,926	2,407	3,959

[1] Revenue from tolls.

Year	*Forth & Clyde*		*Crinan*		*Caledonian*	
	Revenue	*Expenditure*	*Revenue*	*Expenditure*	*Revenue*	*Expenditure*
1836	—	—	1,980	—	2,877	4,117
1837	—	—	1,903	—	2,772	4,283
1838	—	—	1,603	—	2,279	4,636
1839	—	—	—	—	2,707	4,820
1841	—	—	2,428	1,573[2]	—	—
1842	—	—	2,113	1,514[2]	2,910	4,578
1843	—	—	1,890	1,154[2]	2,091	5,767
1844	—	—	2,437	—	925	8,723[2]
1845	—	—	2,200	2,373	86,359[3]	86,359[3]
1846	54,118[4]	18,068[4]	2,549	2,449	73,402[3]	82,982[3]
1847	56,499[4]	18,946[4]	2,655	2,566	50,389[3]	55,587[3]
1848	51,771[4]	20,960[4]	2,626	2,270	53,916[3]	—
1849	54,770[4]	24,535[4]	2,747	2,818	5,238	9,724
1850	—	—	—	—	6,001	12,594
1851	—	—	2,491	2,493	6,711	7,466
1852	44,183[4]	20,720[4]	2,070	2,492	7,909	9,261
1853	48,652[4]	14,224[4]	1,958	1,888	5,888	7,429
1854	51,066[4]	16,200[4]	2,340	2,127	7,280	6,629
1855	51,805[4]	15,771[4]	2,417	1,790	6,643	7,135
1856	47,770[4]	15,503[4]	—	—	6,499	6,550
1857	51,212[4]	14,106[4]	2,289	2,332	5,408	6,152
1858	49,000[4]	14,434[4]	2,098	2,188	5,560	6,185
1859	51,309[4]	14,244[4]	2,238	2,149	5,080	6,951
1860	52,774[4]	14,996[4]	774	962	5,622	6,137
1861	52,508[4]	15,399[4]	3,099	2,938	8,485	7,570
1862	—	—	3,785	2,780	7,604	6,438
1863	—	—	3,780	2,997	6,999	6,240
1864	—	—	3,605	—	6,871	6,449
1865	—	—	3,450	3,661	5,657	6,555
1866	84,451	—	3,375	3,707	6,195	6,047
1867	—	—	3,375	4,324	6,541	6,698
1868	87,145	—	4,350	4,143	6,041	6,560
1869	—	—	4,315	4,394	7,004	6,634
1870	—	—	4,032	5,152	6,943	6,306
1871	—	—	4,814	4,122	7,826	6,526
1872	—	—	5,032	4,416	6,645	6,427
1873	—	—	4,613	4,727	6,316	6,057
1874	—	—	4,759	4,273	6,345	6,547
1875	—	—	4,835	4,153	6,070	7,064
1876	65,309	—	5,057	4,341	6,741	9,307
1877	—	—	5,389	4,293	7,564	7,311
1878	57,128	—	5,966	4,380	8,254	10,511
1879	—	—	5,729	4,928	7,355	10,489
1880	—	—	—	—	8,811	9,793
1881	—	—	4,103	6,226	7,498	7,726

[2] Excluding major repairs. [3] With government grants covering major repairs.
[4] April to September.

Year	Forth & Clyde		Crinan		Caledonian	
	Revenue	*Expenditure*	*Revenue*	*Expenditure*	*Revenue*	*Expenditure*
1882	—	—	5,547	5,978	8,207	11,342
1883	—	—	5,529	6,098	7,784	7,447
1884	56,049	21,279	—	—	8,292	7,447
1885	—	—	5,567	5,829	7,793	7,659
1886	43,605	—	5,146	4,689	6,890	7,168
1887	—	—	4,854	7,343	7,229	7,602
1888	44,038	—	4,814	4,610	6,748	10,507
1889	—	—	5,129	4,954	7,236	10,407
1890	—	—	5,166	4,770	6,351	10,536
1891	—	—	5,271	5,024	7,530	10,001
1892	—	—	5,444	5,358	7,547	10,067
1893	—	19,159	5,416	5,094	7,697	11,093
1894	42,601	—	5,308	5,064	6,930	7,356
1895	—	—	5,143	6,652	6,963	7,073
1896	40,142	—	5,214	4,961	7,535	7,035
1897	—	—	5,394	4,999	7,530	7,441
1898	38,281	—	6,178	5,798	8,594	7,454
1899	—	—	6,161	6,090	9,417	7,571
1900	—	—	6,062	5,907	9,570	7,351
1901	—	—	5,640	5,562	8,280	9,027
1902	—	—	6,028	6,062	8,070	7,574
1903	—	—	6,239	5,590	7,766	7,506
1904	41,155	22,609	6,370	6,031	7,747	7,284
1905	—	—	6,441	5,625	8,102	7,357
1906	40,108	—	6,283	7,533	7,642	7,581
1907	—	—	6,208	6,310	8,215	8,228
1908	—	—	5,882	6,352	8,542	7,548
1909	—	—	5,809	6,677	9,730	8,963
1910	—	—	4,981	5,545	9,284	10,800
1911	—	—	6,151	5,850	10,827	10,100
1912	—	—	5,390	5,486	10,686	9,724
1913	35,136	—	5,708	5,581	9,770	10,882
1914	—	—	5,778	5,344	10,007	9,620
1915	—	—	5,153	4,894	9,669	10,947
1916	—	—	3,775	4,621	7,946	9,390
1917	—	—	3,734	4,265	6,124	8,508
1918	—	—	4,202	5,266	6,374	11,498
1919	—	—	6,887	5,717	6,331	17,338
1920	—	—	5,667	7,932	9,036	22,360
1921	—	—	8,544	8,736	—	—
1922	—	—	10,509	10,868	—	—
1923	—	—	8,873	16,406	—	—
1924	—	—	9,815	13,866	—	—
1925	13,434	—	9,505	10,974	—	—
1938	16,645	23,849	—	—	—	—
1942	—	—	—	—	8,350	31,989
1951	27,838	—	9,559	20,871	9,488	37,188
1952	28,327	—	10,115	23,807	9,267	43,778
1953	27,584	—	10,459	22,458	9,475	40,525
1954	—	—	10,108	22,691	8,353	41,837
1955	29,545	65,286	10,049	22,888	9,934	42,694
1956	36,456	99,291	10,298	25,915	9,872	58,121

Note: Figures for the Crinan from 1851 and for the Caledonian throughout are for the twelve months ending on 30 April of the year named.

C. *Dividends (Southern Canals) per cent*

Year	*Monkland*	*Paisley*	*Forth & Clyde*	*Union*
1800	—	—	10	—
1817	72[1]	—	25	—
1829	—	—	27	2½[4]
1834	—	7½[2]	25	—
1838	—	7½[2]	—	5[4]
1839	—	3[2]	30	—
1841	—	—	8¾[3]	7[4]
1864	—	6[2]	7[3]	—

[1] On original shares.
[2] To obligants.
[3] On increased stock.
[4] On shares freed from debt.

INDEX